A SILENT JOY

The diaries of an Italian hill farm

ISBN: 094396952

Cover Design by Beth Leonard

Originally published in Italy by Franco Del Moro/Ellin Selae

Published in the United States by:
Blackberry Books
617 East Neck Road
Nobleboro, Maine 04555
USA
Gulf of Maine Bioregion

chimfarm@gwi.net

The crackle of dry branches underfoot
The buzzard's arc in the sky
The forest's ants are my witnesses:
the order of this world
instills me
with silent joy.

Conversation Beneath the Trees
Janis Baltvilks, 1998

Thanks

My life has been so full of gifts that I feel this story is not just mine, but I would like to thank Peter Lindley, Paolo Buttiglieri and Giuseppe Moretti for saying "Write!" and all my neighbours for having shared this place and its stories with me over the years. Thank-you also to all the friends who have been part of Pratale and who have generously woven strands of their lives into ours and left their own stories here. I thank Gary Lawless of Blackberry Books. It is wonderful to have a poet for a publisher. I especially thank Martin, who has shared the joy every day.

INTRODUCTION
Giuseppe Moretti

In the early nineteen seventies, in northern California - on the city outskirts, down in the valleys and along the mountain ranges - there were people who liked to call themselves the inhabitants of Shasta Bioregion. Shasta: after Mount Shasta. Bioregion: an area of the Earth with natural boundaries.

They were mystics, revolutionaries, ex-flower power people, poets...but they were also teachers, architects, carpenters, fishermen...and their vision held out an alternative to mainstream culture, based as it is on the exploitation of man by man and nature by man. They proposed a new/ancient mode of living together on Earth: a mode which respected ecological processes, the water systems and the geological peculiarities of place, and which upheld a sense of social reciprocity and place-appropriate lifestyles.

It was a quiet revolution, making no big stir in the world, indeed almost invisible, but carried forward with determination and awareness: a process of renewing local communities and economies, of cultivating healthy food for oneself, clearing water courses, freeing them from the city asphalt, cleaning up the ancient Salmon habitats, the habitat of Bear and Spotted Owl and, above all, of rediscovering one's place in the Big Flow of the Universe, a Universe which, they realised, began right under their feet and all around them.

They were not visiting Martians, just ordinary people working for deep, long-term social change - in education, in the economy, in the workplace, in politics, throughout the whole of society - a change in tune with the natural cycles of the various bioregions. And they were not alone...many other people in the United States, in Canada, Europe, South America and Oceania felt the same urgency and these people, from many different cultures, races and places of origin, *still* constitute the most determined and consistent movement for the reinhabitation of the earth. They work by rescripting themselves and their activities, their culture, politics and lifestyle within and with respect for the wider web of life: grass, rock, cloud, rain, the plant world and the animals as well as the human community.

Etain Addey is one of them, in spirit and in her daily life, and this book/diary is the faithful account of what it means to reinhabit one's self and one's place, omitting neither the practicalities nor the feelings the experience calls forth. This process of "becoming native again" is not an easy task today, in a world where life is increasingly standardized and enslaved by technology, light-years away from the "Dream of the Earth". It takes constancy, determination, a great deal of work and much humility...but the effort is rewarded by joy and wonder when one finally plunges one's hands into the manifold substances of the earth.

The stories in this book are told with a pleasing combination of cultural maturity and sophistication which takes its matter from the humus of the earth. They let us see the world for what it is - a whole: body and mind and nature at one. But the work that Etain and her family do at Pratale - it is important to say this - is not the false idyll of a life "in harmony with nature", but a life lived in direct dependency on nature, in a very everyday, earthbound way. They cultivate the tree for its good fruit but they also revel in the exuberance of wild nature because that is the only way to become worthy citizens of this earth again.

Po River Bioregion, 12 July 2002

For Martin, Melissa, Beniamino and Camilla and "all the Pratale people" human, animal and vegetable and for this place that sustains us.

At the Solstice, the Sun pauses in his journey:
a critical turning point, a moment for reflection.
The Winter Solstice is the "door of the Gods", the journey
toward the light. The Summer Solstice is the "door of men",
the beginning of the infernal journey.

Winter Solstice
Recognizing the wild god in Vallingegno

At the head of the beautiful, wooded valley in the Umbrian hills of central Italy where I live, there is an eleventh century abbey, the Abbadia di Vallingegno. At the time when St. Francis, fleeing from Assisi, asked for asylum here, there was already a large community of Benedictine monks. But below the level of the present abbey are the remains of a pagan temple dedicated to the god Genio Silvanus - god of the forests, of the cleared fields, of shepherds and of human fertility.

The poet Gary Snyder says: "Find your place on the planet and dig in" and this valley is my place. Thus I find myself in the human situation so common nowadays of being an invader and having to make a subtle effort of imagination to align myself with the spirit of the place I have chosen to inhabit. The first step in this process of claiming ancestors is to recognize that a "spirit of place" as such exists, that it is not just a poetic convention.

When I first came here, I asked my farming neighbours where the name of the valley came from, but no-one could tell me. This is a phenomenon I have noticed before: I remember asking local people about a small stone circle hidden in a thick copse on the Isle of Purbeck in the south of England and people living thirty metres from the nearest standing stone had never heard of it - or else perhaps didn't want to talk about it. It seems as if there is a perception of some danger involved in focussing directly on what the powerful collective human spirit has converged upon in the past. It took me long solitary reflection as I worked on repointing the stone walls of the old house to remember the small dusty bit of cardboard I'd seen at the back of the abbey with its curt and enigmatic information: "This abbey rises on the site of a temple to the god Genio". So this valley is the valley of the god, the valle of Genio, the name unchanged for perhaps more than two thousand years but its origin sunk out of present consciousness.

The local people can afford not to know consciously because they hold in their minds the archaic layers of their prehistory and history, and their language, customs, animals, plants and landscape are formed from the same mind-stuff. Their tradition is unbroken and even a peasant neighbour who has never heard of the ancient Osco-Umbrian language or read the contents of the bronze Eugubbian tablets (c. 200 BC), has a bodily understanding of Vallingegno and its environs which I can only come at by going back to basics.

There seems to be a process whereby the land moulds bodies. The European people who migrated to America began to acquire the skeletal proportions of native American Indian within two generations even where no intermarriage had occurred. Perhaps the land can mould minds too. The native Australian people insist that nobody can invade land and remain inviolate because the spirits of the dead ancestors live in the land and they may be born to the invaders as children. Jung points to this claim as a great psychological truth. The earth is what informs our unconscious mind, and no-one can live in a place without that place living in them, taking hold of them.

Here is some comfort for those of us who wish to reinhabit the earth. We at least have the advantage of actively seeking to become native. We are not an invading army committing genocide, bound to imitate and embody the people we have killed, but aspiring natives asking to be taken in, deliberately opening ourselves to the place.

In Vallingegno there is an unbroken line in the day-to-day living. The valley of our river, the Chiascio, has paleolithic sites; Gubbio has a bronze age settlement; some of our neighbours have Etruscan names like Ceremonia and faces identical to those on precious Etruscan vases. Linguistic experts say they can still trace the river bed of the Osco-Umbrian language in the speech of those on this side of the Tiber and Etruscan in the dialect of those on the opposite bank. The family names of the farmers living around Vallingegno are the same as those recorded in the medieval annals of the castle which rises on a small hill between our land and the abbey.

Snyder has a poem in which he tries to reply to his little sons' question "Who are we?" He has not got the gift of a whole and unbroken tradition because of the American genocide. Who are the children of an American of European origin and a Japanese woman who nevertheless live in the Sierra and farm in the old way? In *What happened here before* Snyder gives his answer, and it's a work of imagination. He wants to be part of a true line of ancestors and he claims the rocks, the plants, the animals, the Indians ("And human people came with basket hats, winter houses underground and nets, yew bows painted green…") and even the marauding pioneers. What actually connects his family to the place in the end is their need for sustenance:

their drying the apples, their working and dreaming in that place.

This attempt to grasp "what happened here before" is fundamental. The spirit of a place is not just what is here, but the process: what evolved here and how. So I am concerned with two aspects -firstly with the psychic history of Vallingegno and secondly with its present-day inhabitants, human, animal and plant.

Here there are certain trees and plants that are so native nothing ever hurts their hold on the place: the bright yellow broom, *ginestra,* which we use for lighting fires in the winter, the wild plums and cherries which propagate themselves ceaselessly, the oaks which spring up everywhere and dwarf the Forestry's attempts at introducing foreign pines, the wild chicory and clematis that people gather for food in the spring, the wild roses from which we gather hips for tea. The olive trees have no doubt been here for many centuries, but I feel they are only a step or two ahead of us because they still need helping and tending.

The daily vicinity of these plant inhabitants, having to do with them, is a way in for us. The complex layers of human culture are almost too sophisticated for us as a way of belonging and require too much basic environmental experience which our neighbours have and we haven't. Every time I ask Rosa up the hill a question, out flows a stream of very specific knowledge: exactly which sort of wild broom to cut for baskets, on which Saint's day, from which part of the woods, how to strip it, the basket styles for different purposes... In one lifetime we can only absorb a certain amount, learn some of the mushrooms, some of the cures for the sheep ills, some of the old ways of cooking, some of the agricultural proverbs ("When it rains in August, it rains oil, honey and wine-must"), but the continual interaction with the plants and animals is a more immediate dimension than the recipes and techniques.

Writing in 1930, Jung said "(Primitive man's) landscape is neither geographical nor geological nor political; it contains his mythology and his religion and all his thoughts and feelings, because he doesn't know that they are *his* thoughts and feelings." I have a sense now that we may soon be able to stop talking in terms of projection and realise that perhaps our landscape, when it really is ours, *is* our thoughts and feelings after all.

The biologist Rupert Sheldrake talks about the possibility of there being an invisible physical "field" that he calls a morphogenetic field which is what informs matter, and many native peoples have a concept of field. The Ainu talk about a bear field which is the place where the bears have a habitat, formed over the many centuries of bear habitation. Sheldrake's field does not refer to place, but perhaps there may also be a spatial effect, a strengthening of certain bodies and behaviours in one place and a weakening .in another? I think it is a possibility that the human field may actually in some physical way line up with the fields of certain plants and animals, and over many eons, perhaps with a place in its wholeness. I can imagine the human field beginning to merge with that of the olive or vine here, so that the flourishing of the plant is not only due to the pruning and hoeing and manuring but to an identity of intent and habitat, a mutual understanding.

Some might think this doesn't matter, but I think it is important for us to feel recognized by our neighbouring plants, not to feel we are the only aware organisms doing all the recognizing. This is obviously what native American Indian peoples felt - that their land recognized them - but how could they say this to white invaders except to talk about their fathers' graves?

At the level of human images as manifestations of the spirit of place, Vallingegno has an extraordinary line of resonances, which I've gradually come to see over these last twelve years of farming here. Genio Silvanus was originally a Pan-type God, a God of vegetation and fertility and later representing the willingness of powerful nature to concede space to human culture. Thus Silvanus was the God of shepherds, and cleared fields. He watched over the work of clearing land for pastures and over tree-growing. He is shown as a bearded wild man wearing a goatskin (rather more of a human figure than half-goat Pan) and holding a pruning-hook, identical to the one we still use here today.

This is a strong image of wilderness cooperating with human cultivation: the collective human spirit conceived of or perceived a wild God who accepted pruned trees and sown fields in the clearings of his forest and who gave the protection of "wild" health to the crops. Silvanus was also concerned with the well-being of flocks and with human fertility. His attributes were the dogwood, oak, fig and bean, the wolf and the woodpecker. His long hair and goatskin, and the wolf, woodpecker, oak and dogwood speak of his wild nature, but his human form and the pruning-hook, as well as the fig and bean bespeak his cultivated aspect. In the mosaic I have of Silvanus, a wolf sits at his side as if he were a domestic dog; an image of peaceful cooperation between nature and human life.

Over the centuries, the original significance which we deduce from Silvanus' image must have changed. He began to turn into a threatening figure. The Etruscan bronze sheep's liver which was discovered at Piacenza has sixteen areas, each one denoting a different god. In divination by entrails, signs in the area signifying Silvanus were always considered to be an inauspicious omen, like the areas representing the infernal gods. The forest was becoming hell, an Underworld.

The Romans evolved a ritual to protect women in labour and new-born children from his malefic influence. During childbirth, three men guarded the door of the house against the wild god who would try to get in and steal the child. One shook an axe, one a broom and one a pestle, the accoutrements of three domestic divinities, Intercidona, Deverra and Pilumnus. The axe keeps the fields clear from the encroaching woods, the broom keeps the farmyard cleared and the pestle, the instrument of transformation which turns crops into food, represents the human home. Thus Silvanus had become the enemy of human culture, a figure to be feared. His significance as a God of human fertility had become the danger of overwhelming passion and the disorder of nature. It all speaks of the agricultural fight against wilderness.
Presumably the idea of a God entering a house through the door reflected the idea of a soul entering a body at birth, and the new child is a symbol of re-awakening forces and resonated with the wild strength of the nature god.

In Vallingegno today, as in many parts of the world, there is far too much chopping down of trees. The changing image of Silvanus highlights the origins of this attitude to the forests. This ritual of fending off the wild God must be the beginning of the widespread European/Christian tradition of the battle for the baby's soul at the moment of birth, traditionally fought between the Devil and the Good Angel. There are many folktales in which this scene is reenacted, and the best retelling I've ever read is in Flann O'Brien's "At-Swim-Two-Birds", a completely surrealist story in which the Pooka (the Devil) and the Good Fairy travel across Ireland to attend a birth and the Good Fairy loses the child's soul over a game of cards to the Pooka. In many

Italian fairy stories, the devil is still "the wild man from the woods".

In other words, what we in our Christian tradition see as the battle between Good and Evil started out as the battle between human culture and wild nature. As the Pan-type gods became identified with the Prince of Darkness, human culture took on the significance of the Good. Where does this leave us? We have difficult inner work now, letting go of this deeply rooted perception of Nature as Evil. How are we in Vallingegno to get back to a new vision of the original face of Genio Silvanus? How can there be a reconciliation? How did the spirit evolve here?

In the fifth century, a French knight on a pilgrimage to Rome came through Gubbio and found it already a Christian city with its own bishop. Out here in the countryside, however, there was still a very strong pagan community. Indeed, pagan means "country dweller": it is easier for the gods of nature to hold the hearts and minds of people living in the midst of woods. When the knight Verecondo preached the new religion in Vallingegno, the country people warned him not to anger the forest God. When he ignored them, the pagans stoned him to death. The bones of this martyr lie under the abbey. It is recorded that San Verecondo "freed Vallingegno of wolves with his spiritual authority". His work was not to do with real wolves, I suspect, but with the pagans. It was in this sense that Saint Patrick was said to have "rid Ireland of snakes". Our neighbour has often repeated to me: "If you're not baptized, you're an animal!"

In 1206 or so, Francis of Assisi came to the abbey asking for shelter after he had left Assisi and his father's fury. The Benedictine monks were extremely reluctant to take him in: he must have looked wild. However, he was allowed to work in the kitchens and here at the abbey he changed the water into wine, in imitation of the first miracle of Christ. By now Vallingegno was Christian. The abbey was still surrounded, as the temple had been, by thick oak woods and no doubt there were wild boars as there still are, and wolves. St. Francis stayed in the abbey on many later occasions and after he received the stigmata, he used to ride across the valley on a donkey. The local people warned him against riding through the woods at night and it was in this valley that he made his famous gesture of reconciliation with the wolf of Gubbio.

The story says that a "very great wolf, terrible and fierce, which not only devoured animals but also men and women" appeared in the woods around Gubbio. St. Francis said that he would go forth and meet this wolf, and ignoring the protestations of the people, he went out and found the animal. He conversed quietly with him and brought him back into the town and made a pact between the wolf and the citizens: they would feed him and he would live peaceably.

The pact was an agreement to recognize needs. Perhaps St. Francis could see the importance of honouring the space taken by nature as well as the human space. "Hear ye, my brethren. Friar Wolf, who is here before you, hath promised and sworn fealty to me that he will make peace with you and never more offend you in anything; do ye now promise him to give him every day that whereof he hath need; and I become a surety unto you for him that he will faithfully observe this covenant of peace". This trust between human society and the surrounding wilderness is what the Christian community had begun to lose - the sense that nature gives human culture space as long as human beings recognize the wilderness and honour its primacy. The wolf "went so gently through their city" until his death two years later, that he was greatly mourned by the citizens of Gubbio, who buried him in Church land.

When I first found a picture of Genio Silvanus, I was amazed to see another wild man with a large wolf sitting at his feet, just like all the images of St. Francis and the wolf of Gubbio. I suddenly saw St. Francis as the direct spiritual descendant of Silvanus, and Vallingegno as the powerful receptacle for this timeless wild man dressed in rags in the forest talking quietly with the wolf.

So in Vallingegno, the first image in the human psyche is the forest God, then a barefoot saint, and now, over the last sixty years, an ordinary man who took to the woods in the forties and who lived there until three years ago. He dressed in plastic bags - the goatskin and sackcloth of our century - and had long hair and beard. He was a reticent, quiet man, appearing rarely and talking occasionally with shepherds. He had got a woman pregnant when he was a young man and when she shut herself up in a convent, he climbed a tree outside her window and lamented for five days. She never looked out, and he never entered a house again. He regarded his life in the forest as a life-long penance, and saw himself as "outside Christendom", an animal. He lived on wild food and once said sadly to a woman neighbour of mine that the squirrels and mice had stolen his store of acorns. "Come to the houses" she begged "No-one would refuse you food." "If I don't work, I won't eat", was his reply. Mothers scared their children "Arriva Stocchi!" with this living embodiment of the wild man from the woods.

These woods we are so intent on chopping down are still part of the old forest that was cleared for the temple - the realm of the psyche, the place of testing, the place of re-birth. The green woodpecker of Silvanus was a prophetic bird with magic powers and a protector of trees. In Christian symbolism, the woodpecker is Satan, heresy undermining belief and tempting human nature. The wolf was an expression of valour and of the power of instinct until Christianity turned him into a spoiler of the flock, the

Devil, another symbol of heresy and craftiness. The only place in Christian iconography where the wolf loses this image is an emblem of St. Francis. The dog accompanies the Mother Goddesses and huntresses, and the Good Shepherd and thus we see he partakes of the original, positive nature of the wolf.

Perhaps the wolf in becoming dog made a gesture towards human culture which we need to reciprocate by recognizing our own wild-man nature. Tame man is at war with nature; wild man was not. Jung speaks of the animal as standing for the non-human psyche, the world of subhuman instincts and for the areas of the psyche which are unconscious. Thus the figure of a man with an animal -Silvanus or St. Francis with his wolf - is an image of the unity of man and wild nature, the self undivided, the union of conscious and unconscious mind, the world at peace with itself.

I can only speak about my place, from my attempt to make it truly mine but I feel that wherever one decides to dig in, whether in the birth place or away from it, the archaic layers which are part of the spirit of that place hold all the clues for a return to health. In the day-to-day work in the place and the turning of a receptive mind to the local genii, are open arms for the returning native. The silent joy that wells up turns into words in an attempt to tell the stories. They are only words, but I know that all the other people who live intimately with a place will recognize the source.

January
Winter Wood

"Why go to all that trouble?" It was the middle of winter, on a cold, threatening day and I was trying to get a heap of wood sawed up. This silent question hindered me. It came from all three donkeys, their noses dangerously close to the moving saw. They had the patient, contemplative look of a friend who wants to keep you company but feels you're wasting your time here. Not only did I have to watch out for the velvet nostrils of these donkeys, but I also had to put up with the ironic gaze of the five cats who were lined up on the donkeys' backs. This family of cats have been riding the donkeys now for a long time, a talent they learnt from an ancestor of theirs, one Gatto Mimi, who discovered this warm symbiotic relationship with equines twenty years ago.

I could tell snow was on the way (from the news on the radio but also from my cold breasts, always an omen) and I didn't want to be caught out without a large pile of logs sawed up the right size for the woodstove. But when the donkeys, with their morbid fascination for human activity, came at me with their silent question, I started losing my grip on the saw. Quotes came to my mind, as if I were thinking it through on their behalf. "Consider the birds of the air, they sow not neither do they reap...." (St Matthew) or "We could live on this earth without clothes or tools" (the poet Gary Snyder).

Then I said to myself that I come from a long line of hardworking Yorkshire women and I got on with the job. The wood smelt wonderful: we still had old olive branches from the exceptional frost of 1985 which killed off dozens of trees in Umbria and Toscana. The wood smells like incense, and so does the red juniper, which some claim is hallucinogenic. The colours and the flowing veins of the wood were a delight and the stacking time was a sensual experience.

Behind me I could hear the quiet sound of twenty-three sheep chewing the twigs of the piled up branches and the noise of the lambs playing. Every year the new lambs jump onto the thick plastic sheet from the old earthquake tent which covers the dry wood and discover what fun it is to slide down it in single file. I was surrounded by all the domestic animals. It reminds me of the scene described by my sister, who is married to Michael from the Caribbean island of Grenada: there, when someone wants to move house, they call all their friends, pull up the wooden house by its struts and carry it on their shoulders to the new site. But it's not only friends and relations who come: *everyone* comes because it's not just work, it's where the noise and singing is, the eating and dancing, that's where everybody's gathering and nobody wants to miss out.

For someone who spends most of his time eating grass, sawing logs is an interesting activity to observe. Human beings are, in any case, an endless source of entertainment because they never tire of *doing things:* they run backwards and forwards with axes, hoes, wheelbarrows, hosepipes, they laugh, shout at each other, stamp their feet and moan, hand out long explanations to each other and are hardly ever content with just *being*. They seem to find no peace in this world and the minute one piece of frantic activity is over, they quickly think up another.

When Martin started, years ago, to work with the donkey Otello, they began on the logs that needed dragging out of the forest. Led by a rope, Otello kept at it at a relaxed and easy pace, backwards and forwards between the forest and the house. Then suddenly, after a couple of hours, he sat down, as if to say "Tea Break" So Martin, who agreed, lay down too with his back against Otello's belly and there they both sat in the winter sun, for a quarter of an hour. Then Otello, with a deep sigh over his part in human activity, got to his feet again as if it say "Come on then, let's get on with it!"

This last year, Martin has begun to use the young horse for carrying wood. Otello is rather old and he only carries lighter burdens now, and in particular burdens he has some personal interest in, like grapes and apples. But when he sees Martin go by with Papu, he has a look of nostalgia about him, a mixture of relief at being pensioned off and regret for the end of a long working relationship.

Meanwhile I had made a huge pile of logs. I put the saw down and picked up the axe. The donkeys followed me to the chopping block as if to try their luck with the axe blows. And suddenly, there was a new obstacle in my path: my son Beniamino. He had just become an apprentice violin-maker and now I found out what this means for our stores of winter wood. I raised the axe. Ben came out of the gate and shouted "No!" I lowered the axe. "What?" I said. "What is this, poplar?" asked Ben. "Yes" said I. "When was it cut?" asked Ben. "Oh, years ago. It's a piece of that beam from the old house: it might be eighty years old." "Well, then, mama, are you *mad*? I need it, for making violins." "Right, so take it away from here". I picked up a large log of olivewood and raised the axe over it. The donkeys' nostrils flared, inches from the axe. "Stop!" It was Ben again. "You want this one too?" "No, not me, but Paul is looking for good bits of olive". Paul is a furniture restorer in London and when he heard from my daughter that she was brought up in a house with olivewood doors, I think he nearly fainted. But it is no indication of fine living, simply the result of the abundance of dead olive trees in the years after the earthquake when we were rebuilding. We used what we had.

Finally I found a piece of ordinary oak, with no takers, and split it in four.

February
Gathering rosehips

This is the rosehip gathering season: the sky is bright blue and the sun shines despite the biting cold air. I've already spent several days wandering across the hills gathering and my arms and hands are covered in blood. There's a good pile of red berries spread out to dry in the end room, together with baskets of lichens for dyeing and the remaining golden persimmons.

I set off downhill, through the bottom olive grove, and the dog Pissiru comes with me. Every day we go through the same routine: I begin picking and Pissiru looks at me intently, wagging her tail. The hips are ripe and sweet, delicious...I give her one and she eats it and continues to wag her tail. I give her another five or six. Finally she takes the last berry round the back of a bush and I see her spit it out. She's had enough but doesn't wish to disappoint me by refusing. So now I can get on with filling the bag slung over my shoulder. It's a long, slow job and I want to pick about twenty kilos.

We use them to make tea. Every day, I pick two or three kilos and then I leave them about a month and a half, spread out on an old sheet, to dry. When the hips are hard, I mill them and then comes the fun part: how do you get those little itchy hairs out? If you eat the whole rosehip, the hairs catch in your throat. Round here they call the fruit *grattaculi,* itchyarse, because they make themselves felt *wherever* they end up. The first year I gathered rosehips, I didn't know what to do about the hairs. I though maybe they'd come out if I put the dry, milled hips through a sieve. But what happened is that they rolled into little fluffy balls and stayed on the surface, which is fine because you can just skim them off. However, at the end of several hours, I discovered that all my clothes were full of hairs, a fully functional penitent's vest. I washed and rewashed them, but in the end I had to throw them all away. Nowadays I do the job naked, hiding behind the courtyard wall and praying for no visitors to come.

Every rosehip bush is unique: they are individuals just like us. Each one has berries of a slightly different shape: some are rounder, some are longer, smaller, larger, some are prickly as if the plant has confused the hips with the branches, which are all armed with savage thorns. They are at different stages of ripeness, and I think this depends on their whereabouts on the hill. A lot of hips have beak-shaped tears in them: they are an important source of food for birds in the winter and so I have invented two rules for sharing the crop: I never pick up hips that fall to the ground (they are for the creatures that can't fly) and I only pick about a third of the hips on one bush. Birds are territorial: if I strip every bush in the area, they can't just go next door. I can't be sure how many of us here are relying on this source of food, so when I've picked my share, I tie a piece of coloured wool on the bush so that I remember I've picked there and as a sort of decorative votive offering. No other humans here collect these berries and I could pick any amount if I wanted to go further afield.

It's a meditative job, wandering from bush to bush with my eyes full of sunlight and the bright red of the berries, the gold lichens on the branches and the peach-coloured dead oak leaves that never fall off here until the new leaves appear. Every so often I get a glimpse of the horses moving through the bare trees. I know the stallion loves rosehips and is a rival picker because I often find long white hairs from his mane festooned on the branches, as if he had adopted my system of votive offerings. Today, after a couple of hours picking, when the mind and body are quietened, I notice there's a robin actually following me from bush to bush, one step behind me all the way. He must want the company as well as the food.

I remember the Russian folktale about Masha and the Bear. "From bush to bush, from tree to tree went Masha in the forest, until she was quite lost" It's very easy to lose your sense of time and place when you gather wild fruit. You find yourself on an inner path, the reflection of the wild world whose paths twist and turn and fade out, appearing and disappearing in the undergrowth. In the story, Masha ends up a prisoner of the Bear, but he treats her with gentle respect. It is interesting to see that there is an almost identical story amongst the native American Indians, in which a girl out picking berries meets a Black Bear who takes her to his home in the mountains and marries her. Their children are half man/half bear. These stories are about our kinship with the rest of creation. These gatherings are what tie us to the wider mammalian community...we have all been gathering food together for millennia and all the other animals know the rules of good behaviour: if you watch a sheep eating, you'll see that it never eats a whole plant, but always leaves part of it.

Today, as I fill my bag, I am turning over in my mind the writings of Stan Rowe, a deep ecologist from Canada, who has this to say: "If you want to know how something works, look at the mechanism at lower organisational levels, but if you want to know *what its purpose is, what it's for,* you have to look at the next level *up* of organisation, you have to ask what it is *part of*". What is a cell for? To maintain healthy tissue. What is the tissue for? To preserve the health of the organ (liver, heart, bark). What is the organ for? To preserve the health of the organism (the bird, the man, the tree). What is the organism for? And Stan Rowe replies "To preserve the health of the planetary ecosystem, which is a higher entity which includes the organisms and without which they cannot live." It is difficult and painful for human beings to recognise that we are not the most important entity, but if we want to find our place again in the web of life, we are going to have to embrace limits. We need to treat this entity, this matrix in which we live, as if it were more important than ourselves, and treat all our relations - those who creep, those who run on four legs, those who fly, those who swim and those with roots - with good manners.

As I leave the last bush, the robin lands on one of its flying top branches and pecks at the soft, scarlet pulp.

March
Antonio's stories

"Oh, he beat that animal nearly senseless, once, years back...yes, and that bull never forgot it and now, you see what's happened...animals are just like us, they remember who's been kind to them and who hasn't." This is old Antonio, explaining to us the death, early this morning, of a man further down the valley. The man went by himself into the field where he kept his bull, at dawn, and the bull trampled him to death. All the neighbours are shocked by this violent, archaic death.

For us, Antonio is the valley's memory. Over these twenty years, he has shared his memories in many stories, especially with Martin, giving him what he would have had from a father, if he had been born here of a farming family. Antonio is a mine of agricultural knowledge, a source of useful proverbs, someone who knows how the past explains the present. He is 86 now and spends many hours just sitting by the fireside, helping Rosa to sort the wild chicory she's picked, but he has a sharp mind, and if he tells us the same story over again, it's because that's how stories are learnt, and because he feels acutely the need to make us aware of what happened here before, to connect our lives with the lives of the people from the old times.

And those old times were quite something - some of his tales would merit a Boccaccio or a Chaucer for their telling! At first, Martin wasn't so attentive to the stories because what was on his mind then was sheer survival here in the Umbrian countryside, with three small children. His education as a Swiss accountant was no help in dealing with a moribund sheep or pruning olives. Antonio was already a bit too old to send up an olive tree, even though he was still shouldering fifty-kilo sacks of barley and dealing with the barrels of wine down in his cellar. But Martin thought he could ask him to come down to our farm for a few days and act as an olive-pruning consultant. So here is the scene, one day in March many years ago: Martin is standing on an olive branch, saw in hand, while Antonio, sitting comfortably in a chair under the tree, points out where to cut and shouts up explanations. The sheep are gathered round, pleased at that season to have fresh olive bark and leaves to eat. (One of the local proverbs about sheep says "March - hold me up, I'm so hungry I can hardly stand!"; there is a saying for each month of the sheep-year). Seen from the kitchen window, it was a funny scene, but in just a few days, Antonio managed to put Martin through an intensive course in pruning, scattered with memorable proverbs "The olive should be poor in leaf and rich in muck!" or "If it pisses down in April, you'll be needing a big oil-jar", "By the Assumption, the olive is oily" and, most wonderful of all, "When it rains in August, it rains oil, honey and wine-must".

He and Rosa were kindly, thoughtful teachers, who tried to give us advice without making us feel incompetent. Once, as we passed by their fields in November, they called out "Today we're planting the broad beans!" and they meant "It is time to get planting the broad beans", which traditionally go in on the Day of the Dead. Broad beans have a very ancient association with the souls of the dead, which they were assumed to contain.

After a while, Martin stopped asking theoretical questions like "What are the main diseases of sheep?" because he realised that, faced with that sort of question, Antonio would reply "Have another glass of wine!" A question like that seemed meaningless to him. But when Martin came up to see him one night, upset and worried, and said "Our ram's eyes have gone all white, as if he's suddenly blind! He's stumbling and falling all over the place....what can we do?" Antonio said "Ah, that's what we call Moon Disease. What you have to do, Martino, is flick some sugar into his eyes...you'll see, he'll be better in no time!" Martin felt doubtful, it seemed such an extraordinary thing to do, but then Antonio had kept animals for so many years and had a reputation for being good with them, indeed he'd been born on the day of Saint Anthony Abate, patron saint of animals. So Martin tried this bizarre recipe and sure enough, it worked and the ram was cured.

Once we were no longer complete novices and our hands had assumed the proper callused and grimy surface, we felt we had more time to sit and listen to Rosa and Antonio's stories of the valley. Thus we heard how Salvatore spent the Second World War hidden in a hollow oak tree with a barrel of wine because he didn't want to kill other people and how the priest reported him to the authorities. "And the moral of *that* is, when you go to Confession, you can say the small things out loud, but the big things, you have to say them *under your breath*, otherwise..."

Then there was the drama which had once taken place in *our* farmhouse: sometime in the forties, a lad from the nearby village of Scritto had helped his young lover to poison her husband. She had done the deed over a period of weeks, by lacing his morning milk with a little dash of rat poison. When the young husband finally dropped dead, his mother called to mind how he used to complain every morning that the milk tasted bitter and how the wife would say "Nonsense! Get it down you!" When the autopsy revealed the truth, they arrested the wife and went looking for the lover, who had bought the rat poison over the counter. But he had disappeared off, not, as one might imagine, to America or at least to the other end of the peninsula, but just over the hill here to the farmhouse where we now live. And here the old family hid him for fifteen years, in a walled-up and nicely white-washed fireplace, which indeed my sister Jane uncovered, knocking the false wall down with a push of the foot, when we mended the upper floor. No doubt this family was rewarded for their solidarity with demijohns of wine and sacks of wheat, but they did have one serious problem: they had a number of good-looking, grown-up daughters ("Smart girls they were too, like city girls!" said Antonio). How to protect these girls from their lusty guest? The solution was brilliant. There was a nocturnal marriage ceremony up at the Abbey on the opposite hill, between the lad and one of the girls of the host family, but everyone was masked...the bride, the bridegroom, even the priest. So all was right with God, but no-one else knew anything about it. Then, after fifteen years in hiding, someone gave the poisoner up to the police. There were only two years to go: at that time in Italy, someone who had evaded the Law for seventeen years could no longer be arrested. But it seemed that the man in hiding had been looking a little too insistently at one of the other daughters and so *her* husband had decided that, unless he did something radical, it would be the bitter milk for him too, before long.

In all these stories, Antonio seemed anxious always to underline the good qualities of the people involved, to make us empathise with them, see their humanity....he didn't want us to think badly of anyone, he wanted us to see them as people just like us, living in this place, living in harder times.

The poet Wendell Berry, who gave up a university career and went back to farm the Kentucky land that had been in his family for six generations, has a wonderful poem called *The Record* about an old country neighbour of his who tells him the stories of past times. A young friend of Berry's is so struck by these stories that he begs the poet to go back and record them on tape so that this old way of life won't disappear forever. But Berry refuses to do this and undertakes the far more complex and difficult journey of living alongside the old man and making his knowledge and the old ways he represents a part of his own life. Thus he treats the stories as seeds to bring to flower in his own day-to-day living and his poem ends with his plea to others to listen to the voice of that old man as long as he's there and when he's gone....

"Be the next one like him"

April
God of war, God of spring

The cat has had a little grey and white kitten and she is determined to keep it in a precarious basket on top of the cupboard outside the front door. When I discover the kitten, she is being tended not only by her own mother, Little Betty, but by her great grandmother, Mouse, a very old cat who spent all her younger days living in the sheep shed, and who always smelt comfortably of sheep and hay. Only old age has persuaded her to move indoors where life is more comfortable. Last week, she gave birth too, but things turned out as they often have done in the past: she stared at her newborn kitten with the cruel eyes of Kali and thought "All the bother of bringing it up or a quick meal?" So she ate it, starting at the feet. But now she's in the mood to be a nice granny and so Betty's kitten has a warm nest made of two cat-bodies. The dog Pissirù, who's always been a good mother and who is very fond of kittens, is extremely worried about this situation and has tried various times to send this old witch of a cat away, but in vain. We are worried, too, and every time we hear crying from the basket, we go and see what's going on in there. But so far, all is well and there's another big-bellied cat soon to join the party.

The kitten is only the most domestic sign of Spring: outside the woods are turning green and the cuckoo is calling; the company of goldfinches has come back to nest in the cypress tree - there must be twenty of them this year, the donkey Olivia has a huge belly now and walks more and more slowly through the pastures, the little chestnut-coloured foal is turning white like the other adult Camargue horses, the hillside is full of flower petals falling like snow from the cherry trees, plum-trees, peach-trees and apricot trees, and there is silence in the forest on the opposite side of the valley where three woodcutters spent the winter chain-sawing trees, their curses echoing loudly all day long. When the April clouds disperse, we can see the sun. We can also see the NATO planes flying towards Serbia with their bombs, or returning without them. Whenever we look up to search the skies for the first returning swallows, the eye meets with these bombers, a travesty of Spring energy - a low, threatening drone amongst the other voices of Spring.

Mars, whom we know as the God of War, was originally the God of Spring, the God of the energy that bursts out of the earth and vibrates through the vegetable and animal world: Juno conceived Mars not with Jupiter but by union with a mystical flower. For the Romans, who were farmers before they became conquerors, Mars was at first the God of Agriculture. He was called Marte Gravidus, from the verb "grandiri" to grow, and his festival was the Ambarvalia, celebrated in Rome on 29th May with the sacrifice of a boar, a ram and a bull - it was all about the *fertilising* male energy! Only later, as the focus of attention in the Roman world turned from the natural world to the world of politics and conquest, did Mars come to represent, in the consciousness of mankind, the bloody conflict between peoples.

It's true that I have begun to see that the emerging lifeforce in Spring is intimately connected with violent death. This great explosion of birth and growth is full of blood, wounds, accidents, reciprocal eating. An entomologist friend once told me that if all the offspring of one pair of flies should survive for an entire season, the whole planet would be two yards deep in flies: such is the great role of Death in Life! (Who on earth *does* these calculations?)

This river of sheep's milk, this *pecorino* cheese and *ricotta* depend on the death of lambs. The buzzard never flies overhead without a company of small birds who dive-bomb her to keep her away from their nests: they are afraid she will eat their nestlings and they are right. The pretty ladybirds that crawl all over the garden are waiting for the bean plants to grow so that they can start massacring aphids. It was only after a few years of country life that I realised that many of the blood-splattered events amongst wild and domestic animals are centred on Spring, that season the poets sing and that we think of as the season of flowers and sunny days.

It almost feels to me as if I had provided an apology for those bombers...as if they were a natural phenomenon like the buzzard's beak, or just an expression of the testosterone curve that rises in the human male as if does in all the other male mammals at this time of year. Spring has always existed and so has War.

But perhaps this desire for war is exasperated by our self-exclusion from the natural world? Our urban, industrial life prevents most of us from participating consciously in the Spring explosion. How many people live in the wild world? Most of us are shut away in houses, offices, hospitals, schools or shops and we don't feel the heat of the sun, or the sting of the rain, the strength of the wind, we don't see the birth of all those other beings, or watch the gradual budding and flowering going on outside, we miss out of the drama of life as well as the drama of death which accompanies it. We are cut off from it all, our dramas are virtual. Do these bombers represent our mechanised, impersonal and alienating attempt to participate in the exuberance of Spring?

Martin suddenly calls from the olive grove. "Quick! Come and see this!" I run down, thinking of all the accidents that might have happened. But he's found a nest full of tiny goldfinches in an olive-tree he's about to prune. The minute birds are snuggled together, their little golden feathers already showing. The cypress tree must be full to overflowing this year, some birds have had to nest in the surrounding olives. Ben, Camilla and I take it in turns to climb the ladder and peer into the nest. Then Martin takes the ladder and the pruning-saw. "I'll leave this tree 'til last: they'll have flown by the time I've finished." he says.

As we go back up the hill, two low-flying bombers roar over our heads.

May
The shearing

It is the end of May, the woods on the opposite hill are a sea of green and the sheep and donkeys are sunk to their flanks in high, flowery grass. It's already hot and the fruit on the mulberry tree in the courtyard has begun to fall, lying soft and sweet on the warm bricks. It is time to shear the sheep and we're starting on Camilla's birthday - today she is sixteen.

When Camilla was born, Martin's parents gave us a lamb - a suitably rural present for a spring birth. We only had five milk-sheep then and we went over to the other side of the valley to pick a lamb from a large flock of Sardinian sheep. The shepherd told us proudly that this lamb's sire was an English ram, a Sussex. We called her Elderflower because this is the time of year when the elder near the old house is full of white flowers with their narcotic perfume.

We dress in rags, because this job is dusty, woolly and full of sweat and lanolin, and go down to the sheepshed. Martin gets the shears out and sharpens both pairs. We decide to shear Elderflower first because she is the oldest sheep we have now - sixteen is a good age for a sheep - and she walks slowly, rheumatically. She is hard put to follow the flock nowadays but we can see she still enjoys the sun and the pasture. The sheep who gets shorn first feels cooler and moves more easily in this early summer heat, so that the rest of the flock, still weighed down with their fleece, are forced to leave the shade of the field maples and follow her.

We haul Elderflower out, lay her on the ground and tie her legs together. She doesn't struggle much, she's been shorn many times and isn't frightened. I start at the thigh and shear in strips from the neck towards the tail. The tail is a separate matter and so is the neck...I don't like doing the neck much, the shears are very sharp and sometimes a sheep will suddenly turn its head to see what you're doing. Indeed there are some sheep who act like a customer at the hairdresser's: they have a look in their eye as if to say they don't think much of the style and they keep raising their head nervously and checking to see if you are still as bad at it as they suspected.

Once you have done half the fleece, up to the spine, you roll the sheep over and start from the other side, working up towards the top again. At the end, the whole fleece comes free and you untie the legs and shear the sheep's belly, checking for ticks or wounds, and then let her go: and the sheep struggles to her feet and runs for it, bleating and miserying, back to the others, who refuse to recognise her and butt her as she tries to get close to them.

But when we let Elderflower go, she can't get up: she's been lying in the same position for too long and her legs hurt her. We have to help her stand, and for the rest of the morning, every time we go into the shed to fetch another sheep, we find her lying on her side with her eyes shut, as if she were about to die. We feel bad. It would be terrible if she were to die today, on Camilla's birthday, and all because of our shearing operation. We carry on shearing, in the shade of the lilac bushes, but every half an hour one of us goes back to heave Elderflower into a sitting position.

The shears we use seem to have stayed exactly the same since the Bronze Age - they are one of those brilliant inventions that no-one has ever improved upon. The first time I saw this venerable tool was when I had first come to the country and got roped in to help Tonino and Luigi shear their two hundred and ten sheep, but I had seen it in seventeenth century prints and even in Roman mosaics.

There had been twelve men and women from around the valley at that first shearing session, twenty years ago, plus an extra hand to catch and tie the sheep from the fold for the shearers as they were ready for another customer and a woman who did nothing but pass the beer bottles around: it was an occasion for telling good stories. I remember one of these stories was about a man called Gambalesto, which means "Sprightlyleg", who was a widower with several small children. He needed another wife and so he sought out the *ruffiana,* the person who arranges marriages, who at that time was a local woman called Maria. Maria said she wanted half a million lire, which in the sixties was quite a price. But the terrible thing was that this Maria, having introduced Gambalesta to a prospective wife that he didn't like, insisted on keeping her half million, claiming that he had paid for a *wife*, not for a *choice*. People were still laughing over that memory when someone else told another story about the same Maria, who had by the time of that shearing left the country to go and live in town. Maria was obviously a fey woman with the money: this time she was owing several men for doing some building work. They had waited a good, long time for their pay and finally came down to her farmhouse. Maria knew her husband was a jealous man and he wasn't too far away across the nearby field. So she bellowed at the top of her voice that the man were grabbing her, whereupon the men fled as Maria's husband, a large man, came storming across the ploughed field.

With one story after another of this kind and plenty of beer, we finished the shearing by lunch and went down to a feast consisting of every known kind of roast meat, prepared by Tonino's wife in vast quantities. I had a very old Landrover, so Tonino asked me to carry the wool back down to the barn. I had only just learnt to drive and it was a tricky experience to manoeuvre down the precipitous dirt track with every centimetre of space in the back and in the cabin packed with dirty sheep's wool so that I could hardly change gear or steer. (It was the second time I had faced new-driver challenges in the country - once in blinding rain, as I climbed another very steep track, three grimy, wet wood-cutters had leapt out of the forest in front of me, delighted to find someone to take them back down to Artemio's house and I had had to do a sixteen-point turn on the edge of the precipice with several chain-saws impeding my new skills).

While we ate our way through all the meat and home-made pasta, various neighbours told me approvingly how surprised they were to see that I *could* shear a sheep. I had been the slowest, but I was pleased not to look as if I couldn't learn.

Today, when we finish the last of our twenty-two sheep and let them out of the shed, we also set the three weaning lambs free. They have been separated for two months and they are so happy to be out on the pasture and eating grass that's growing instead of grass we cut for them that they hardly realise the rest of the flock is nearby. Finally they look up and see their naked mothers and aunts. All the sheep feel strange and there's a lot of alarmed butting and skirmishing. Elderflower keeps falling over and we have to keep an eye on her for the rest of the day, but by the next day she's found her feet again and the flock has calmed down.

Saturday, Tonino comes by and asks if we can come and help him finish shearing: he still has ninety sheep to do. We go over to his place and spend a morning shearing in the shade of the big oak tree outside the huge prefab which has served as a sheepshed ever since the earthquake. It is sad to compare the day with that first shearing twenty years ago because now there are only three of us left - Tonino, Martin and me: all the others are dead or much too old or have left for the city. There's still plenty of beer and good humour and the lunch is as meatily festive as ever, but where are all the people who could be here learning these old ways and having fun?

June
Lucia

"Uncle Gervaso died last night!" called my neighbour Ivo from his window as I went by. "No, really?" I replied "I'm so sorry. I thought he was immortal!" "Nearly," said Ivo, "he was a hundred and one!"

As I walked on up the hill, I thought about that very old man and his long, long life and wondered whether his longevity was due to his having had an elemental force of nature for a wife. Zio Gervaso was a little, energetic man, all muscle, and he'd played second fiddle all his life to Zia Lucia, like a firefly in the light of the full moon or a priest at the temple of a Goddess.

The first time I saw Zia Lucia, she was still living in the farmhouse above mine. She must have been seventy-five then and she had a large family. Some of them were standing staring from the doorway of their house, which gave straight onto the dirt track, curious to see the foreign woman from town who had bought the lower farm. They only stared for a minute and then they surrounded me and swept me into their kitchen. They poured wine for me, and put on the record player the only record they possessed, which was accordion music, and invited me to dance. I couldn't resist such a festive welcome and I drank and danced until I was out of breath. Lucia's sons danced with amazing verve, spinning me all round the kitchen and when I had had enough, I went and stood in front of the wide fireplace where Lucia was cooking.

She was a tall woman, with no spare flesh on her but still attractive and with a finely sculpted face, very beautiful. She seemed to be wearing several layers of grey rags, a full skirt and an apron (but no knickers, as she later assured me, that being more convenient for pissing and so on...) but what really struck me was her familiarity with fire. The big fireplace was the only place in that kitchen where one could cook and Lucia stood before it like a juggler. There were at least seven different sorts of food cooking: meat roasting, potato skins for the pigs boiling in a tin hanging from a sooty hook over the fire, *crescia,* which is a local bakestone bread, covered in a layer of ashes and resting on a red-hot stone, tomato sauce simmering in a three-legged pan with its broken handle fixed with binder-twine, hot pasta water keeping warm in a cauldron on one side of the fireplace, and two pans resting directly on the embers, one with chicory and garlic and the other with indecipherable innards. Lucia stirred, poked and mixed, squatting down by the wide hearthstone to sweep the hot embers into place with her bare hands. The fire leapt, rising or retreating under her hands. She laughed, her mouth wide, the lips still full.

I was pregnant at that time. It showed and Lucia made no attempt to pretend I was married by calling me Signora. She yelled good-naturedly at her grand-daughters who were trying to climb onto my lap. "Get off! Can't you see the signorina is expecting a baby?" She stroked my belly. "Sweet!" she said. As I watched her that day, I began to see what a universal sensuality she possessed: she stroked and pinched everyone and everything: men, women, children and animals. She took me down and showed me the big, white *Chianina* cows in the stable under the kitchen. They are meat cows so people never milk them, but Lucia winked at me as she ran her hand over the flanks of the nearest cow. "They *love* me, these cows!" she said, "I milk them and they turn their heads and lick me with those big, warm tongues of theirs". She had been a share-cropper all her life: squeezing every last drop from the land was a necessity, but Lucia squeezed with such faith in the abundance of nature that she seemed to hold the cornucopia in her hands. A neighbouring peasant once warned me, "That Lucia is like a locust across this countryside! She picks everything for miles around, half an hour before it's ripe! Walnuts, cobnuts, chicory, acorns, cornel cherries, maraschino cherries, plums, sloes, blackberries, mushrooms - when we get there, she's already been and gathered it all!"

I discovered that the valley was full of Lucia stories: the one in which she stripped naked during the threshing lunch and circled the men sitting at the long table, her hair hanging loose and festooned with flowers like a rustic Ophelia, pretending to have gone mad just to liven up the proceedings, and then there was the story of Lucia and poor, shy Gervaso, on their wedding night. It seems that after the festivities, when the newly wedded couple shut the bedroom door, Gervaso sat down on the edge of the bed and undid his shoelaces. Then he did them up and undid them again. Lucia was already in bed, waiting for him. He was just tying his shoelaces up for the second time when Lucia's voice rang out loud and clear in the dark: "Gervaso, if you don't get those shoes off *right now*, I'll *SUCK* them off your feet!" There wasn't much privacy in those days of the extended family and the whole household rocked with laughter.

My own relationship with Lucia was a sort of apprenticeship. The next time I came up from Rome, she took me firmly by the arm and led me into the dark recesses of her house, down a long corridor. "I'll show you how to make cheese!" she said. I was grateful to her because she was the only person in my life at that time who saw no reason why I shouldn't turn into a peasant: she took me seriously and wasted no time in starting on my education. I hadn't a single sheep, I wasn't even living here yet, but Lucia threw open a door and ushered me into a little room full of pans of milk. Without washing her hands, or mine, she plunged my hand into the first pan of milk and rennet. "Feel that! Can you feel how it is? It's nearly set. Feel this one over here. This is ready for cutting now. In ten minutes we can start making cheese. Lick you fingers! Good, isn't it? I'll let you taste the ricotta when it's done..." It was brilliant teaching - no amount of explanation about temperature or consistency could have taught me what I learned by running my hand through those warm, half-formed curds. But I did wonder about bacteria: I'd always heard that cheese-making required careful attention to cleanliness. Then I looked back at Lucia and realised that no germ would have the courage to mess with her work!

I remember another afternoon with Lucia: we were sitting in her kitchen and she was telling me about how it was when her children were born. She had had eleven. "All *good* ones!" she exclaimed. Round here, this means they are all alive. "And don't you go listening to those doctors, they don't know anything. Don't eat for two, eat for at least *three!* Because they don't know what they're talking about and often there's more than one baby in there!" "But, Lucia, did you have someone to help you when you gave birth?" I asked, afraid she was going to tell me about lying in ditches. "No, we women helped each other, but once the vet happened to be here...what a *clever* man he was with a cow! Since he was here, he helped me that time." Lucia had a whole theory about babies. She claimed that when they are born, their bones are soft as can be and so you can mould their little faces just the way you like them. She used to gently pinch and pummel all of them, squeezing the girls' little titties ("And milk comes out!") and then the cheeks and lips and noses of all her newborn children so as to make them beautiful. She was sure that the beauty of her grown-up sons and daughters was down to this early sculpting on her part. But what she really loved to discuss was men. She boasted that she'd had every man in Umbria. I couldn't believe that, but I was sure she'd devoured all those she'd come across at least with her eyes. "And those Black Men! In The War!" She raised her eyebrows and laughed. "They aren't MADE for us white women! But weren't they just *sweet*!" I could picture what the war was like for Lucia. Despite the hunger and having to hide the hams and cheeses not only from the landowner but also from the various armies, it sounded from her descriptions as if the Second World War had been fought for the sole purpose of introducing her to the most amazing variety of good-looking young men as they filed past her remote farmhouse. The Afro-American GIs were obviously still vividly in her memory, because once when I came up from Rome with a group of friends, she stopped the car outside her front door, and sticking her head in through the window, murmured to me "Who's THIS?" "He's a friend of ours, Lucia, this is Jean-Pierre." He was from Haiti. "Tell him to get out!" she ordered. "What does this old lady want?" asked Jean-Pierre, laughing. "Be nice, Jean-Pierre, she just wants to look at you!" I said. He got out and Lucia, with a wide smile, walked slowly round him looking at him from all angles as if old memories were flooding back and then went and fetched him a big glass of wine.

Then Lucia's family were turned off the farm: the owners wanted to live there themselves and the peasant family found another farm, this time to rent. It was miles from the main road on another dirt track in the middle of the Gubbio countryside and one spring afternoon I set off on foot to go and see them in their new surroundings. I was seven months pregnant by this time with my first daughter and it was a wonderful, sunny walk, but a long way. When I got there, I found Lucia in bed. The family gave me their usual festive welcome, even though Lucia was only just out of hospital, where she'd been for the first time in her life. No-one could explain what she'd had, and she seemed as strong and as full of energy as ever. The family was obviously having a hard time keeping her in bed. The doctor at the hospital had told her she should eat *in bianco,* which is the classic invalid's diet in Italy, and means no tomato sauce, nothing spicy or with cooked oil, and one of her daughters-in-law was obeying this command by making Lucia a plate of plain boiled rice with a dash of raw olive oil on it. Gervaso took the plate from the girl and came towards me on his way to Lucia's bedroom next door. He stopped at the kitchen cupboard, opened it and hid behind the cupboard door. I watched as he snatched up the black pepper and shook it vigorously over the white rice. He smiled and, turning his back on his daughter-in-law, muttered to me, "Lucia can't live on pap like this!" He was her accomplice, still in love as he neared eighty.

That night I slept with her in the big iron bedstead. Dusk had come suddenly before I realised it was time to set off for home and it would have been midnight by the time I got back. Lucia said, "Sleep here with me, I'll throw Gervaso out!" So she did. It was one of those monumental, rustic beds, with a deep, warm, wool-stuffed mattress. Lucia took the opportunity to pinch me all over, stroke my big belly and feel the baby kicking. I slept soundly, but in the middle of the night I suddenly woke up with a start and to my horror, saw a male face looming over me in the dark. It was the wine-red face of one of Lucia's sons under a straw hat full of holes, well pulled down over his face. His breath was winey and his eyes unfocused and he said "Move over, mamma, so I can get in!" I had a moment of panic: Lucia was so enamoured of the exuberance of the whole world, I was afraid she might really move over for him. But to my great relief she shouted "Get away with you!" and he did as he was told.

At dawn, Lucia jumped out of bed: enough of being an invalid! She turned to me as she reached the door and said "Don't move! Or you'll get morning sickness! I'll get you something to drink so's you won't!"

I wanted to protest that I didn't get morning sickness anymore but Lucia was already back with a large glass of clear liquid. "Drink this!" she said. It was *grappa,* strong, home-made liquor! "Is this how one has eleven children?" I wondered to myself. Faced with her determined look, I drank it. Lucia went off to the kitchen again, calling over her shoulder, "I'm going to get you something to eat now that will do you a power of good!" "Will I manage to get out of bed and stand upright?" I thought, "I *will* have to eat something now!" I looked at the clock: it was six o'clock. Lucia came back after a few minutes with a huge omelette, full of lumps of fat. "These are goose eggs...and look, I've put you plenty of lumps of good pork fat!" It was an awakening which I have never forgotten, I still laugh when I think about it: but it was a female initiation rite in the purest tradition!

Lucia died at ninety-three. She's up there now, for sure, whooping it up and when Zio Gervaso gets there, she'll throw him the sort of welcoming party for which she was famous all around this valley!

Summer Solstice
The earthquake

The signs and portents began many months before that 29th April 1984. At the height of the preceding summer, the two huge poplars which overhung the old farmhouse had both died suddenly. "That's funny" I thought and then, "Not a good omen for this house!" It was only afterwards we realised that the water table must have dropped suddenly, as it does sometimes before an earthquake, leaving the roots of those water-loving trees high and dry. It was sad to see two flourishing trees wither slowly in the sun and we missed the shade on the roof and the whispering sound of their leaves in the breeze.

Then, in mid-April, one of the sheep gave birth at a strange time and place, in the middle of the morning on the field just near the house. Most sheep births happen at night or at least at the close of day and in rather hidden away places. She seemed to be having a hard time, but we were beginners then and we waited too long before helping and the lamb died. After the earthquake, we remembered and wondered if the lamb was premature and the sheep was feeling the subterranean upheavals that were just about to surface.

The night before that terrifying dawn, one of the sheep spoke up as plainly as she knew how. Martin had gone down to milk the few sheep we had at that time. The head sheep, Little Lamb, had a newborn lamb so we weren't milking her but she came right up to Martin and bleated loudly and repeatedly in his ear until he'd finished milking Curly, Clara and Daisy and Snowdrop, as if she had something vital to communicate. "What do you want, Little Lamb?" asked Martin "Don't you feel well? Do you want to be milked as well? Is there an earthquake coming?" He'd guessed right, but Little Lamb could only carry on her insistent bleating, coming right to the door of the shed with him as he left. "Right, right, Little Lamb, goodnight then!" said Martin as he went off with the pan of milk. It was a shame that we hadn't yet read Pliny, who lists "animals talking" as one of the signs of an approaching earthquake or we might have taken that guess more seriously.

If only we'd been all round the neighbours that evening, we'd have realised something strange was going on, because everyone had some sort of warning from the natural world that night. The old exorcist priest from Santa Cristina said to our neighbour Fausta "Look at that blood-red sunset! That's a sky that spells fear and disaster!" A huge owl came and beat its wings against the kitchen windowpane in the dusk up at Orlando's house. He and Gina shouted at it to go away - owls are harbingers of death here - but it came back twice more and beats itself frantically against the glass. "Via, via, bestia!" yelled Orlando in horror. Mansueto's chickens refused to go inside their hencoop that evening.

I was used to the earth moving here. Every so often, waking in the night here, I'd felt that curious sensation of sudden movement and a low booming noise, a huge stirring under the bed, below the floor, below the rock, miles beneath us. So I had it at the back of my mind, always, that one day - or worse still, one night - there might be a big earthquake with collapsing walls and screams.

A deafening noise woke us at first light. Jerked awake, I couldn't at first distinguish all the different sounds that made up that noise: the boom from the earth, the thick stone walls of the house shaking, the tower of the castle on the hill across the valley as it crashed down, the crumbling of our two-storey stone barn, our roof as it caved in, the terrified voices of the animals in the stables below us. A big mirror smashed to the ground near my head, with two distinct noises, one as the wall banged against it and another as it fell loudly onto the brick floor. It didn't break! Martin and I were asleep on a mattress on the floor. I opened my eyes and saw the vast oak beam above my head nearly out of its socket in the stone wall, bricks falling from the roof, the cement between the stone courses of the bedroom walls crumbling. My first reaction was to freeze, because it felt as though, if I got to my feet, I'd just be crushed faster. Everything happened slowly, minds working in slow motion. The seconds went by like hours. "But we're still alive! We should get up and run! But where's Camilla?" She was a small baby, somewhere in the big bed. Can't find her, pull back blankets frantically, pick her up, run through the rain of bricks, pull Melissa and Ben out of bed, bewildered and terrified and run! We were at the furthest point from the front door of the house. We ran through the loud, shaking rooms, down the stairs, out from under the arch and out, towards the valley. Then we turned round.

The shaking stopped. The old skyline was gone! The big stone barn was just an enormous pile of rubble. We stood there, speechless, watching the mushroom cloud of dust rising above it. Martin opened all the doors of the stables and all the sheep, the donkey, the hens and ducks poured out into the courtyard and then ran for the open fields. There was a long minute of complete silence. Then we began to hear our neighbours' voices calling to each other across the valleys, from one hill to the next. "HEY! Are you all right? Oh, can you hear me, down there?" We called back, called the people at the castle and up at Ivo's house. "What's happened up at the castle? Why don't Olga and Francesco call?"

The house hadn't changed much, but it looked to us like a death trap now. Martin and I ran back in to fetch shoes and coats for all of us. We put them on, but no-one moved from the place in the farmyard where we still stood, rooted to the spot. Every few minutes there was another tremor. Melissa said “The people in Ancient Crete said it was a Great Bull moving under the earth!” She was only seven, but she had put her finger right on the heart of the matter : in an earthquake, it doesn’t feel like tectonic plates moving but like being caught up in the personal fury and gigantic wrath of some huge God: it feels as if Someone wants you Dead. Ben, who was only two, stared at the new landscape with wide eyes. He had his own personal mythology: “Mama, Primino’s come and knocked down that barn!” he said wonderingly. And then he added, with the generous heart of small children, “But I’m going to get a hammer and mend it for you!” Primino was the old man in Gubbio who owned the barn, who used to come round every so often and threaten to knock it down, hoping thus to convince us to buy it. In Ben’s imagination, he must have taken on the status of an ogre. Camilla, shaken awake and wrapped in a man’s overcoat, clung to Martin and just stared at the rubble and the dust and the running animals.

It is comforting to remember how people reacted straight after that earthquake. This valley is inhabited by a curious mixture of old peasants, aristocrats and “foreigners” of various styles and persuasions (and you are just as foreign if you come from Bologna or Perugia as if you come from Switzerland), yet after that initial, shocked silence, everyone had the same impulse to run and find other people. We are all scattered around the countryside, and so we got into our cars, in our pyjamas, and looked for each other. At Antonio’s house, there was a small crowd of people standing in the lane listening to the radio. It was the local radio station and they were saying “Don’t worry, it was a big tremor but there’s no damage, just a few tiles off the roofs.” Everyone looked at each other in amazement. We didn’t know yet that the epicentre was right under our feet, in fact under Monte Urbino, and that there was much less damage in Gubbio.

No-one died in that earthquake but five thousand people were left homeless like us and there were many miraculous escapes. In the following days, these miracle stories travelled all around the valley, from one person to the next.

Maria had been up already, in the kitchen, and as soon as the earthquake started, she began running towards the first floor bedroom where her baby Luigi was asleep in his cot. When she reached the door of the bedroom, the opposite wall was just gone! All she could see was the farmyard below, but Luigi's cot was still balanced on the only remaining piece of floor.

Iolanda was sleeping in the top floor of her old house, next to the church. Her husband and sons were safe in the prefab, where they'd been since a previous earthquake two years before. Iolanda was a light sleeper and the minute the booming noise started, she leapt from her bed and started running downstairs. Just as she reached the first floor, the roof collapsed. As she reached the ground floor, the ceiling came down but Iolanda was too fast for it and was out of the front door with no time to spare at all. I saw her two days later: her hair was still standing on end and her mouth was still open.

A young couple from Santa Cristina had come over to visit friends in Valdichiascio that evening but it got late and their friends persuaded them to stay the night. Next morning, they went home to see what the damage was and found the whole house had collapsed, the main beam fallen across what had been their bed.

Saint Ubaldo is not only the patron saint of Gubbio, he is also known as a great protector against earthquakes ("Save yourselves from fire and water, but I will always save you from earthquake!" he is said to have promised in the 12th century) and the day after the big quake, his basilica on the mountain above the town was filled to overflowing with flowers.

That night it rained and we slept in an old tent on the field above the house. There were Martin and I, the three children, all the cats and a litter of kittens, who during the earthquake had somehow had the misfortune to end up in a full potty. As soon as my head touched the pillow, the film of those terrifying seconds began to unreel in front of my eyes...I had heard that this same visual re-living of the event happens to people who've been in a car accident but it was the first time I had experienced it. It was sinister to know that everyone else in the tent was seeing to same film. In the following months, our sensitivity to the continual tremors increased. We were sleeping on the ground, with our ears close to the rock and every time there was a big tremor we would all wake up a second or two before it came and lie there in the dark, waiting. No-one ever spoke until after the tremor. First we'd hear the underground noise and then we'd feel the movement, which was different each time, sometimes up and down, sometimes rotating, or sliding like on a ship at sea, and then we'd hear the sighs and comments of the others. The tent was not much good, it leaked and when we woke up after that first night, we were all clinging together in a tiny space, the bedclothes damp and all the cats sitting on top of us.

In the country, the rubble didn't seem to matter as much as it would have done in town. It's true we had no house, but we were all alive, none of the animals had been hurt or killed, and the fields were all still there, nothing had happened to the grass or the vegetable garden. It seemed almost funny that an earthquake could knock down the castle keep and the barn while the pepper plants that we had transplanted the day before were untouched. It was the beginning of the warm season, we hadn't lost our food supply: altogether we felt we could get on with life. But in the next few weeks we lived through a very strange period, nevertheless.

The government, nervous of criticism after the previous earthquake in Irpinia in 1980, which had been badly mismanaged, was anxious to give an impression of efficiency. Thus we were given a huge military tent which could sleep twelve and then we discovered that there was Earthquake Food! Every day, we could go up to the main road and claim a complete meal each. In those days we were poor and our diet was pasta, lentils and vegetables. We didn't claim our earthquake lunch every day, because it was four km up the lane, but when we did happen to go past the distribution point, we'd come home laden with lasagne, steaks, vegetables, cheese and fruit. At the weekend, friends from Rome came to commiserate with us, bringing cognac and chocolates. So there we were, under forty days of spring rain, sinking into the mud and with the ground shaking day and night, but with a convivial crowd of visitors coming and going and a surreal banquet groaning on our makeshift, outdoor table underneath the elder tree.

The social flow took the edge off the drama, but in our hearts we felt we were living in a different world now: to live close to one of the huge natural forces of nature is an extraordinary experience. To stand in a wood during a big tremor and see all those big oaks and ash trees plunging and heaving, or to sit on a rock amongst the grazing sheep and feel the sudden, curious wheeling around of the earth under your body is something that changes how you feel about the world. When you realise that the nausea and churning you feel in your stomach comes from upheavals that are about to happen several miles underground, that your human hearing can fine-tune enough to hear the low rumbling that precedes a quake and that your eyes begin to pick out the cloud shapes that form before a tremor, it makes you feel physically part of the place, almost as if you were in an altered state of consciousness. Snyder says "Wildness is a state of complete awareness. That's why we need it."

This state of mind is what Dolores LaChapelle describes in her book *Deep Powder Snow*. She recounts a conversation she had with a fellow expert skier after they had skied down an extremely dangerous slope, where the slightest wrong movement could have been fatal. Dolores says "Tim asked me 'How did you feel? I didn't reply immediately and he continued, 'Not frightened, but...' I replied, 'Yes, a total attention. Life and death depended on it'. And suddenly I realised that this is the way most human beings have lived for most of the time we have been on this planet, living fully in a state of total attention towards a place. This is surely what we mean by *being alive*."

This is how I feel about the earthquakes that happen here. They are part of the place, part of its vitality. Despite all the hassle of rebuilding, there's a substrata of ecstasy, of participation, of delight in what nature can do here. One year in which the Po river flooded dramatically, devastating villages and crops, I saw an interviewer on television asking a very old woman if she was angry with the river. The woman stood on her balcony, with the whole first floor of her house under water and replied staunchly "Ah! Don't touch the Po! We love this river! To us it means Life!"

During those first, critical days of homelessness, we lived through a whole comedy show in and around our tent, with a strange convergence of bizarre guests.

In those days, we had a little spare room in the old house that we rented out so as to make some money from holiday-makers who didn't mind the simple life in its extreme form. We were expecting a Quaker lady from London and so the day after the earthquake, we sent her a telegram, warning her not to come because the house was now reduced to rubble. Once the message was sent, we dismissed the matter from our minds, but three days later, at supper- time, we saw an elderly lady approaching down the hill, accompanied by our neighbour Ivo. Ivo looked pretty strained and had a large suitcase in his hand. It was Edith Marshland, who later passed into the annals of Pratale as one of the most extravagant guests, in terms of character, ever encountered.

Later, Ivo told us the story of Edith's arrival in Gubbio. She had caught the Rome to Gubbio bus, but failed to get off at our country bus stop and had got carried on into the main piazza of town. She spoke no Italian, but she had a good, strong voice and so she decided that the most effective approach would be to shout, in the worst Colonial tradition. So she got off the bus, opened her mouth and proceeded to let out a loud yell, hitting a high note and staying with it. People ran to see what the matter was and someone managed to work out that the Signora wanted to get to Valdichiascio. Unfortunately for Ivo, someone had seen him in town and so several people set off to seek him out through all the bars of Gubbio. It was the hour of the *aperitivo.* A small committee cornered poor Ivo in his favourite bar and told him that a lady with a loud voice was screaming and screaming down in the piazza and it seemed she wanted to go where he lived. Ivo made his way down to the bus stop and tried to quiet Edith. Pointing to her suitcase, Edith made it clear she was coming with him and Ivo obliged. When they reached his farm, he pointed down the hill at our house. "Well, young man, pick up my suitcase and take me down there!" commanded Edith. Calling on the reserves of patience learnt in a large family, Ivo walked her right down the path to our little tent on the field and handed her over to us with great relief.

Martin and I and the children were sitting on the ground at a distance from the piles of rubble, by a campfire. How do you receive visitors under those conditions? We jumped to our feet and starting apologising. "We're so sorry, there's been an earthquake. We sent you a telegram...didn't you get it?" "Oh yes, I got the telegram. I came anyway, for the adventure!" declared Edith and then continued, "But I am an old woman. I ought not to get cold. Very likely I shall get bronchitis and DIE. I haven't even got a coat."

We had a good collection of warm, handed down jackets and coats: I went and fetched one for her. Edith put it on and I asked her, "Have you eaten? Shall I make you some pasta? A couple of eggs? A cup of tea?" Cooking was a creative activity then, it meant running into the upstairs kitchen and moving fast, trying to make food between tremors". "*Pasta?*" replied Edith. "I cannot eat wheat, I'm allergic. Eggs are *very* bad for me. Tea? I never drink anything except Lapsang Souchong." I stopped asking questions and went upstairs and made her a tray of whatever came to hand.

When I came out again with the tray, I could see from Martin's face that this lady would not be easy and might very likely give us a harder time than the earthquake. We made her a bed in the other half of the small, leaky tent because we hadn't yet been issued with the splendid Battle of Waterloo tent, and collected as many dry bedclothes as we could find. We five squeezed in together on one side of the tent and explained to Edith that we could not rent her space, we could only offer her a share of the tent and treat her as one of us.

The next morning, Martin announced that he had urgent work on the furthest field and set off with Ben and Melissa in tow. I got out a table, a bucket and a plastic washtub. I put little Camilla in a baby-chair up on the table and prepared to spend the whole morning washing clothes and listening to Edith, because I could see she really wanted to tell me the story of her life. She had tried to tell it to Martin the evening before, but he quickly goes into a trance at such times and even Edith found that discouraging.

Edith started in on the most recent and startling events in her life and worked slowly backwards until she reached the vicissitudes of her youth, when she had been literally abandoned in the African jungle by her husband who was never heard of since. By the time she got to that bit, I sincerely felt for the poor man.

The thing that really rankled now was the disappearance of her most recent lover, a young Arab boy of eighteen. A small age difference of half a century! Edith was obviously a most intense, energetic and passionate person and she explained to me that she had covered this lad with loving attention and that he, ungrateful bastard, had simply gone missing one day. She worked out exactly how much he had cost her: "Seventeen thousand pounds!" "Really!" I said. "Oh yes, if we count the price of the car he took with him!" she replied. "Ah, but I shall get every penny back!" She was certainly the strangest Quaker I'd ever met. "How will you do that if you don't know where he is?" I asked. "Ah!" said Edith triumphantly, "I shoplift regularly from Harrod's, that's how! I have already recovered nine thousand pounds." Harrod's is a very elegant department store in the heart of London, owned by a rich Egyptian family. "But what do you steal?" I asked, amazed. "Oh anything, whole Stilton cheeses, evening gowns, feather boas..." said Edith. "But don't you get caught?" I asked. "Oh yes, but the next time I *disguise* myself...they can't recognise me!" Edith seemed to me one of the most recognisable people I'd ever met. I guess that the owners of Harrod's, seeing this strange old lady, rather aristocratic and eccentric, come shoplifting every day, disguised in clothes she'd stolen the week before, had resigned themselves to her personal crusade against all the Arabs and written off their losses.

While Edith told me this story, she unpacked her enormous suitcase. What does a lady bring with her on a country holiday in April? A country consisting of deep mud and rubble? The suitcase (certainly from Harrod's) was full of nothing but glittering evening dresses, embroidered with gold thread, pearls and sequins and coloured feather boas of every imaginable hue! Next to my heap of muddy country clothes, it was laughable, it looked like the transformation scene from Cinderella. Camilla, in her baby-chair, grabbed a handful of beautiful red feathers and stuffed them into her mouth, delighted.

By lunchtime, when Martin and the children came back, Edith had shared with me a great deal her of resentment against past lovers, friends and relations and she felt better. We ate, sitting on the ground, Edith in a splendid evening gown and a golden ostrich feather boa. And then, that afternoon, two more unexpected guests turned up.

One of them had no legs: when he was an adolescent he had thrown himself under a train but he had survived. I didn't know about this at first. I didn't know either of the visitors, but when Martin saw them walking up the hill with their backpacks, stepping through the rubble as if it were nothing unusual, he jumped up and called. "Urs! Sarah! What a surprise!"

They were two very young members of the housing coop where Martin stayed sometimes when he was back in Zurich. They sat on the field with us and told us all about their journey and exchanged news of home with Martin in Swiss German. Edith studied them suspiciously though half-closed eyes.

When night came, we had to think where they could sleep. "Don't you go putting them in the tent because I hardly get a night's sleep as it is!" Edith warned us. "We could sleep on the floor in the old house" proposed the boy. "But it's dangerous!" objected Martin, "There are tremors all the time and the house is not safe". It wasn't raining and in the end Urs and Sarah decided to sleep outside in their sleeping bags. "But if it starts raining, come into the tent" we said, "better we're all squeezed up than wet!"

In the middle of the night, I woke up and heard the first drops of rain. "Martin!" I whispered.
"What?"
"It's raining."
"So?"
"Those kids are getting wet..."
"Look, Etain, we told them to come in and if it rains, they'll come!"
But the minutes went by and the rain sounded louder. I could hear a voice outside the tent whispering "Sarah!" Sarah!" It was Urs. Why didn't they come? In the end, the rain really began to come down and I went out to look for them in the dark. "Hey, come on, come into the tent!" But Sarah whispered, "Please tell Martin to come and carry Urs: he hasn't got his legs on. He takes them off at night." I looked at the boy. He was so thin and frail I could easily have carried him myself but I didn't want to embarrass him. I ran back inside.
"Martin!"
"What NOW, Etain?"
"Urs hasn't got his legs on and he wants you to carry him into the tent!"
"OH my God" said Martin, feeling really bad now. He jumped up and ran out. Edith, on her side of the tent, was already murmuring furiously because every time we came in or out, we had to climb over her. As Martin emerged from

the tent, the rain redoubled its efforts. It was pouring now and then suddenly there was a big tremor.

"I am going to get ILL here!" I could hear Edith's voice. "I will DIE here and you'll have to ship my CORPSE back to England because I REFUSE to be buried in this Papist country!" Her tone was so angry that the two smallest children, woken by the tremor, started crying. Martin came back in, soaked to the skin, carrying Urs in his arms in his wet sleeping bag and Sarah came behind him, dripping wet as well. There was nowhere to put them except next to the welcoming Edith. Fortunately they couldn't understand much English.

The next morning, I got up early and went down to the old house to make some bread and enjoy the peace of dawn and some solitude. But Edith appeared, still furious. "I haven't slept a wink all night because of those two kids! And you were just as bad! Backwards and forwards all night! It's worse than Africa here. And I can't understand why that boy has to have someone *carry* him around!" "Look, Edith, it's because he has no legs...he lost them in an accident." I explained. "OH!" exclaimed Edith, horrified, "I didn't know THAT! I'll go and tell him I'm sorry...YES, I'M SORRY HE'S GOT NO LEGS!". I grabbed hold of her arm and said "*Sssh! Please*, Edith, don't talk to him about it, he doesn't want everyone to know" But I thought, "And he wanted to sleep in the old house, which could come down at any moment! With no legs! If there had been a tremor, he couldn't even have run away!" It was like living in a madhouse, let alone just being homeless!

The day was full of tremors and visitors. Someone called Pablo turned up. He lived on the other side of the valley and it was clear that he felt scared and lonely with the earthquake shaking the house at all hours of the day and night. He wasn't sure how safe the house was, and he had walked five or six kilometres to come and enjoy some company. The house he lived in was his brother's, but his brother was travelling in India and Pablo didn't know how to contact him and let him know what had happened. He joined the earthquake banquet and we talked but when it got near sunset, the sky went black and a storm seemed to be in the offing. We could hear distant thunder. Pablo jumped up: "I must go or I'll get caught in this storm. I haven't even got a coat. Could you lend me something?" I went to the pile of clothes in the entrance of the old house, picked up a coat at random and handed it to him. Pablo set off up the hill. Then I turned round and saw Edith's expression.

She was a mask of rage. "REMEMBER what you've done to me!" she said. "What, Edith?" "REMEMBER that you took THE COAT OFF MY BACK and gave it to a young man who doesn't need it!" "Oh Edith, I'm *sorry,* I didn't realise that was the coat I gave you! But we've got lots of warm coats, I'll go and get you another one". But it was no good, she'd never forgive me.

That night during supper, it poured with rain and we ate inside the entrance of the house, a place that was easy to escape from if there should be a big tremor. Edith held forth on the subject on the black population of London. She protested against racism, but at the same time, she said "Huh! Don't tell me! I know them ONLY too well - you can't trust them!" The two Swiss kids could hardly follow the English and they stared at her, fascinated, trying to catch her meaning. She looked like the witch in a fairy tale, with her glittering eyes, loose hair, deep frown, her green feather boa flying as she shook her finger at them. "AH! You don't agree! Come on! Speak up, then! I know you've got it in for me! You're on their side! The blacks! YOU can't be trusted any more than they can!" "Edith, Urs and Sarah don't speak English" said Martin, "They can't understand what you're saying!" - "Ah, don't they now? I don't believe that. They understand only too well! They think they're clever! But I understand their silence!"

That night, in the darkness of the tent, squeezed between our sleeping children, Martin whispered in my ear. "Look, if this lady doesn't change her attitude tomorrow, we'll have to send her away! This is not an easy moment for us and we can't be dealing with a madwoman like her as well!" "She's just a little eccentric" I replied. "Just eccentric!!" exclaimed Martin, scathing.

But the next morning, I was the one who sent her away. I was in the stable milking the sheep and Martin was cooking breakfast over the fire. While I milked, I could hear Edith complaining to Martin. About the coat, about the attitude of the Swiss kids, the bread made of wheat, the lack of Chinese tea, the smallness of the tent, the weather, the black people of London, her offspring who had done her out of the shop she ran, her husband who had abandoned her in Africa and that bad Arab boy with the seventeen thousand pounds, and then again about the coat...." And if I get ill, you'll have me ON YOUR CONSCIENCE FOREVER!" Martin listened in silence.

Suddenly I couldn't stand it anymore. I came running out of the stable like a wild animal. The new milk spilled onto the straw, the sheep ran away, those I'd not milked yet as well as those I had. A hundred times worse than a witch, I screamed "MARTIN! Put this woman in the jeep, take her up to Mengara and leave her in the middle of the road where someone will RUN HER OVER!"

Quietly and firmly, Martin got the lady into the jeep with her Harrod's suitcase full of glitter and drove her off to Mengara. When the jeep had disappeared, the two Swiss kids crept out of the tent, their eyes as big as saucers. I was sitting on the ground, the blood pounding in my head. Sarah came and sat next to me. After a bit, she said, timidly, “I zink ze best is forget all!”

She was so right.

July
The fox!

We have such a heavy clay soil that many centuries ago anyone with any ambition round here gave up trying to dig it and started making majolica jugs and plates instead. But I am not a potter and so I've had to find ways to deal with the fact that if it rains, the summer garden gets a hard crust on the surface, and if it's hot and sunny, it makes an even harder crust. Industrious gardeners hoe the beds patiently over and over again, but I have now got a system worked out. When all the little plants - tomatoes, aubergines, peppers, beans and sweetcorn - are about a foot high, I weed them and hoe them one last time and then I mulch them with a thick layer of old sheepshit right up to the edges of the beds and then cover that with a cushionlike layer of new straw. They look beautifully tucked up - no weeds grow and the earth stays soft and moist. All I have to do after that is gloat.

Well, not quite. I still have to water the garden in the evenings, tie up a hundred tomato plants, fight the brambles which seem able to throw immensely long, thick branches over the garden fence in one night, dig up the potato patch, gather in the onions and garlic, and anyway, this business with the sheep shit and straw is a long job. Every morning, after milking the sheep, when it is still cool before the sun comes over the eastern hill, I choose a bed and start marching backwards and forwards with buckets of manure. I have a muck pile, jealously defended against any possible competitors, next to the donkeys' stables. Each day I hoe the surface of this pile and during the hot hours of the day, the hens, in a spirit of cooperation, work the heap over with their feet, so that each morning I find a layer of perfect, crumbly manure ready for spreading.

As usual, the path between the muck-heap and the garden is fraught with obstacles. All the donkeys get under my feet, especially the baby, who thinks the whole point of humans is they play and tries to get his head between my legs. I don't get offended, I know this is what he does to his mother, and that he hasn't yet realised that we humans are bipeds and so our equilibrium is more precarious. When I reach the garden gate (which I have to shut religiously between journeys so as not to tempt the sheep), I find Otello, head donkey, standing right in my way. Otello moves aside very, very slowly to let me pass, and only after deeply inhaling the smell from the buckets: "Mmmmm! Shit. From sheep. Mmm. What does she want *this* for?" After six or seven trips with the manure, I start carrying straw. This is more pleasant, but the straw stable is also where we store wheat, oats and barley, so every trip is made more difficult by the presence, just outside the door, of the donkeys and hens who hope I might hand out a second breakfast.

When I have finished, I contemplate the beauty of the bed of golden straw with the small, green pepper plants showing through: they look wonderful!

But after a few days on this task, I start to realise that the hens are falling down on the crumbling job: they are definitely working more slowly than they were at first. This confirms my awful suspicion that there are fewer hens coming out of the henhouse when I open the door in the morning. I count them several times and say to Martin "You know, I'm afraid we have lost two or three hens in the last few days." Another week goes by and there is no doubt: of the thirty or so ducks and hens we had before, there are only twenty left. We always shut them up at night, so this means that there's a fox coming and picking off chickens in the daytime. This is a disaster, the hens are free to wander where they like and if the fox is so bold as to hunt them by day, it means it's a vixen with cubs. Every five or six years it seems to be our turn to contribute to the raising of four cubs in the nearby woods. I remember it was our neighbour Gina's turn a couple of years ago and she'd said to me, "The fox has stolen seventeen hens, one after the other!" When a fox has cubs, she becomes very brave and her tactic is to choose a hen house and empty it.

Martin and I have an ambiguous attitude to this situation. We hate to lose chickens, but we are also secretly pleased to think of the vixen with her cubs flourishing in the wood. That is why we are so slow to admit to ourselves that we have been targeted. But by now the fox is right under our noses. Camilla saw her one hot afternoon, then Martin caught sight of her and then I saw her as well, running like the wind towards the edge of the wood. She always comes at the same time - between three and four o'clock in the afternoon, when she knows the humans are usually asleep.

We take defensive measures: we cut the long grass on the hillside above the woods in the direction of the henhouse, we move the horses onto the pasture which the fox has to cross, and we start having our siesta down under the plum trees near the henhouse. One day it's the men's go and the next day it's the women's - a hen party! We set up a hammock and chairs and carry down a pan of tea. Every so often one of us yells "We can see you!" in the direction of the invisible fox in the woods and all the dogs jump up, barking helpfully. A week has gone by and no more hens have disappeared. The four orphan ducklings waddle round without the worried look they had before and the cock seems less nervous. Nobody mentions the secret worry: what are those cubs eating? If only we had news of lost geese from some neighbour or other! "Una volta per uno non fa male a nessuno" as they say round here ("if we take it in turns, no-one gets hurt"). Meanwhile, we humans have been telling each other fox stories.

A Sicilian friend told us a lovely story about a fox who lived near his farm. His neighbour was complaining that a fox was eating the ripening grapes in his vineyard. It hid in a sort of tunnel, a covered ditch between the wheat field and the vines and they could see the fox but not reach her. "I'm going to smoke that fox out!" said the neighbour, ripping up a handful of dry grass. He made a little pile of dry grasses at the entrance to the tunnel and got out his matches. But as he struck the match, the fox put out her paw and scattered the heap of grass. The farmer swore under his breath and made a new pile, but once again the fox knocked the heap of grass away. With man and fox almost nose to nose, this scene was repeated until the farmer ran out of matches and had to give up.

One of our neighbours told us how he once saw a fox in the midst of his sheep, right down at the bottom of a steep field. The fox was lying on his back, with his paws in the air, playing like a puppy and the lambs were coming closer and closer, curious at this unknown animal. The neighbour started running fast down the path, shouting and cursing, and by the time he reached the field, the fox had gone. Thus he was never quite sure of the truth of the matter: was the fox really playing, or was it a crafty trick to catch a lamb?

By now the afternoon anti-fox siestas have turned into a sort of happening, with foxy stories from all latitudes. In the evening dusk, while I water the garden, I look towards the darkening woods and hope that the fox family has found something else for supper.

August
Luigi comes to the valley

We're sitting on the ground, in the shade of the big oak tree by the path, eating bread and cheese and drinking red wine straight from the bottle: the cheese is our own *pecorino* and the wine's from the vineyard, but the bread is the saltless, white bread from Gubbio: it's the only day in the year that we eat it, because it is part of the ritual.

Today is the anniversary of Luigi's death. He was a guide to country living for me, a friend with a fine mind and a lover. The son we had together was only three when Luigi died, far from here, in his village in Sardinia, and we - the hidden part of his family - have evolved this way of remembering him on the day of his death. Some years it is just the family with a few close friends who knew Luigi, but very often, at this time of year, there are lots of people. We always invite everyone because it's not a solemn occasion, but a sharing of histories. Sometimes we just eat and hold the stories in our hearts, but some years, specially when there are old friends, we tell the stories about Luigi, partly for the pleasure of the remembering and partly for Ben's sake so he'll know.

This year Dave is here, and Neil, two friends who over the years spent many hours with Luigi, talking or sitting in an amicable silence, while his sheep grazed in the valley.

Dave cuts slices of bread and hands them round while he reminisces. "Once Luigi woke me up at dawn. I was nice and warm in my sleeping bag, upstairs in the old house and he whispered in my ear 'Dev! DEV! Come and help me kill a sheep.' I felt bad, really *pretty* bad. I'm never at my best in the morning as it is...and I'd been vegetarian for years! When I got down into the kitchen, you were there, Etain, and you said 'Dave - you don't have to go - tell him you don't want to come!' But I had a feeling that this was an opportunity to face something I was scared of with somebody I admired and trusted. So we went down to the stable under Luigi's house and he pulled the sheep he was after outside. He had two hundred and ten sheep and he recognised all their faces! He laid the sheep on the ground and said to me ' Can you hold it?' He wanted to know whether he needed to tie it up or not. 'Yes' I replied, with my heart in my mouth. The sheep seemed quite relaxed. Luigi got out a knife and cut the sheep's throat and then immediately turned its head back, breaking the neck. It was so quick, and the sheep was dead. No struggle or terror, just a thorough knowledge of how to do it that seemed full of respect for the animal that had to die. It was a revelation for me, I was pleased I had gone along, I felt relieved to know that it is possible to kill with respect like that." Dave raised the bottle of wine and drank a toast to Luigi.

Luigi had told me how he had come from his mountain village to Umbria as a boy of twenty, all alone on the ship with two hundred sheep, and then on the train to Assisi. "When I was on the train, a lady came and asked me for some milk, for her little girl. And I milked a sheep and gave it her." What a different place Italy must have been in those days, with women who would ask such a thing and trains made up of both passenger carriages and cattle trucks full of sheep!

But Luigi only told me the reason for this emigration after we had become lovers. He came over to my house late one dark, November night. He had drunk a good deal and while he talked, I realised the drinking was deliberate, a way of overcoming his natural reticence: he felt it was a sort of duty, now, to tell me this story. Sitting in front of the hearth in the kitchen of the old house, with only the flames for light, the images of his story seemed to come from a lost world. I can still hear his voice, with its careful, staccato,

Sardinian diction. "They said my brother had killed a man." That was the eldest brother, the eldest of four brothers. Luigi was the youngest. It was a long story, of sheep disappearing in the night, families at war, honour, the long, sleepless hours of watching the flocks at night, discussions in lowered voices amongst the older members of the family. At the end of the story, I asked a ridiculously English question: "But - did he kill him or not?" Luigi looked at me, and then, trying to throw a rope across the abyss that divided his world from mine, he replied "He did not kill him. But in Sardinia, at that time, when they stole your sheep, you killed." I tried to imagine the pain of his family as they took this step of sending away such a young son, alone, after the elder son was imprisoned. They wanted to avoid his living with any suspicion of involvement, because he had been present that night, guarding the flock. Luigi, at the age of twenty, had never been out of his mountains and it must have been hard to leave home with no family or friends and go to "foreign" country and live among strangers.

But despite his serious, diffident character and the great workload from such a big flock of milk sheep, Luigi's life in Umbria was full of wonderful episodes and he told these stories with self-deprecating irony and an eye for detail. He was a good worker and as he moved around the countryside with the grazing sheep, he often stopped at one farmhouse or another and it was his habit to join in with whatever work his neighbours happened to be doing, for the pleasure of their company and so as to be part of their concerns. He would disappear at intervals to "turn the sheep" and then reappear. His shoulder bag, woven back home by his stepmother in grey and white wool, always provided enough *pecorino* cheese, roast lamb, wine and white bread for any company he met. It was talk he wanted, he was a bit of a Mercury, and although he was discreet and kind-hearted, he was the one who carried the news from one part of the valley to the next. So he was loved and wherever he appeared, with his dogs and his flock, he was welcomed, as an extra hand and a storyteller. Apart from news of the neighbours, he was the first to know when the first cherries were ripe down at the Chiascio river, or when the mushrooms appeared in the woods, and he would bring them with him, tied up in a jumper, and empty them onto peoples' kitchen tables, little heaps of the wild harvest.

Some of his stories were like scenes from the *Decamerone*. When he was about twenty-three, Luigi pastured his sheep in some fields he rented in Petrignano di Assisi. He had a horse which he rode to reach the fields. In the fields abutting his, there was the daughter of a local farmer who had a small flock of *nostrale* sheep. They were too far apart to be able to speak to each other and for many months they just looked at each other. She was young, maybe a year or two younger than Luigi and by dint of watching each other in the distance through the morning mist and the hot hours of noon, with only the sheep bells for company, a great desire to meet grew on them. "But didn't you ever say anything to her?" I asked. "No. But she looked at me plenty." Luigi was shy and reserved and he couldn't make the first move. Months went by until one summer evening, towards milking time, as Luigi was about to take his flock home, she made the move. He was just putting the reins back on the horse, when he suddenly found the girl standing right next to him, her arms full of washing. "I have to go down to the river to wash these things. You go that way - will you give me a ride on your horse?" Imagine how a boy of that age would feel, after months of longing, when he found the girl sitting in front of him on the horse! She had her arms full of clothes, so he had to hold onto her. When they got to the river, they were both so inflamed they hardly had time to hide in the bushes by the riverbank before they started making love. Months and months of stifled passion exploded. But after a few minutes, Luigi raised his head and saw, through the leaves, a figure approaching across the fields: it was the girl's father, following the trail of dropped clothes which had given the game away. The father had surely understood the significance of those clothes lying in the horses' tracks and he was red in the face with fury as he advanced towards the river with an armful of gathered clothes. "What did you do?" I asked. "I took the horse and rode off!" "That wasn't very gentlemanly of you!" I said. "He would have killed me!" said Luigi. I wonder if the girl had time to get down to the river and pretend to be washing clothes?

Later on, in the valley of the Chiascio River, love presented itself again in an even more Chaucerian version. There were a couple of peasants, Amabilia and Mansueto, who were famous for their fights. (Their names are so splendid because they imply amiability and docility!) When they tired of arguing, they would set about each other and beat each other mercilessly. If Luigi happened to be down at the river, he would hear these hand-to- hand combats and keep his distance. This evidence of conjugal love scared him and probably dissuaded him from wishing to be a husband himself. "In the end, Amabilia used to lock Mansueto in the granary" Luigi told me. "How do you know?" I asked. "I would hear him yelling for her to let him out!" "And what did she do?" "She'd go off into the bottom field so as not to hear him." Luigi would hide, but you can't stay out of sight for long with a big flock of sheep if someone really wants to find you, and on one of those occasions, while Mansueto kicked at the granary door, Amabilia found Luigi, took him by the hand and led him quietly into the middle of a field of sunflowers. It must have been a relief to her to find a quiet man!

We finish our lunch under the oak tree, laughing at the old stories, celebrating a life that will always be part of this place.

September
An African story at the grape harvest

Today we picked the grapes. We were lucky - there was a hot sun and plenty of friends here to help. The vineyard is still young: we planted it the year that the government offered subsidies for grubbing up vines. We thought about what happened to beer in England once the Industrial Revolution got going and people left the countryside. Once there were no small farmers left brewing their own beer, the price of beer was left in the hands of the big breweries and so was the quality. So we planted a small vineyard, with two hundred vines, enough to keep the household in good, home-made wine. Events proved us right: the first year that we made our own wine, the price of wine in Italy suddenly rose.

When we planted the vineyard, we knew nothing about vines, so we went and asked our neighbours for their advice. They gave us plenty, but there was very little consensus. Gino said "Whatever you do, don't plant Merlot!" Others said "What you want is Ciliegiolo" "No - Trebbiano" "Trebbiano? You must be joking!" Baldino said, "Just make sure you plant plenty of Merlot." We thought it better not to rely on one variety of grape, so we planted several: Barbera, Merlot, Ciliegiolo, Trebbiano, Moscatello and a few plants of strawberry grape, Fragolino. Sure enough, when one vine does badly, because of rain or mist or bugs, we're pleased that we haven't lost a whole year's wine. Our only serious problem is the good, red Merlot, which ripens much earlier than all the other grapes, so that we always have to do two separate wine harvests. Otherwise, the wasps would finish the Merlot before the other varieties were ready for picking.

So today we just picked the Merlot lines. Martin brought the donkey Otello down with the antique pack-saddle we hang the crates of grapes on. This year, Otello was in a state of great impatience, because this grape harvest happened in a week when he had urgent business elsewhere. Both his wives were on heat at once and he didn't feel he had time to stand around idle, tied to a tree while people told stories and cracked jokes and picked grapes - as far as he could see - *very* slowly. He huffed and snorted and wound himself five times around the trunk of his tree until he was almost garrotted. Martin called him every so often "Hey, Otello, calm down! We'll be going up in a minute!" When we had the first two crates full, it needed two people to hang them onto the pack-saddle, one on each side of the donkey, so that the weight was equal on both sides. Martin set off up the path for the cellar and he had to run to keep up with Otello, while we all watched in horror at the heaps of grapes bouncing up and down, as if they were about to scatter all over the olive grove. Martin emptied the crates into the wooden funnel in the cellar, balanced over the big tub, and cranked the handle to mangle the grapes through it and then he came back down. He said he was only too pleased to have Otello in the mood to run instead of going at his usual reluctant pace. "But I've had to shut Olivia and Titania up in the stable, or he'd have mounted them, crates and all!".

While we cut the bunches, we got to telling stories of drunken misdemeanours of the past and a Jamaican friend Nimy told me a wonderful story centring on drink. Back in London, she had made friends with an African woman who had moved to England much against her will, brought there by her new husband. Her marriage to this older man had been arranged by her family: she had had no say in either the marriage or the subsequent move to London and had been very homesick at first. However, after a while, she had had two children and had begun to get used to her new life. She found a job and, with her new found financial independence, she started to think about leaving her husband. But at that point, he got cancer and so she stayed and looked after him until his death a year later.

And now there was a serious problem: where she came from, custom demanded that a young widow should carry out a lengthy purification ritual so as to rid herself of the spirit of her dead husband. If she remarried without having observed this rule, she risked seeing her next husband become paralysed and die as soon as he touched her on their wedding night, "because the dead man's spirit was still with the bride". It sounds psychologically plausible! So this woman had gone to some trouble to carry out the necessary ritual, which consisted of her staying shut up in her room for fifteen days, completely naked, eating no food but drinking plenty of gin, in the company of an old neighbour woman who looked after her while she sat and drank gin and wept. Well, she managed to find an old woman from her own country who would come and keep her company and friends to look after her children while their mother got drunk, and the rite began. Of course, back in Africa, it would not have been gin, but some local alcohol, but as my friend pointed out, gin was a good choice for a substitute because it makes the drinker melancholy and sentimental!

At this point in the story, there was a momentary drama when one of the children got stung by a wasp, angry at so much competition for the grapes, so there was an interval in the story while we organised a vinegar antidote, kisses and biscuits.

"So?" we asked, "Did your friend manage to dissolve her feelings in the mixture of gin and tears?" "Oh, yes" said Nimy, "But at the end of the fifteen days, there was another problem. Because back home, she would have gone down to the sea, naked, and completed the purification by bathing in the sea. A final washing away of involvement in the watery element! But her native village was five hundred metres from the sea, whereas she now lived in the middle of London, in a flat, far from any suitable body of water. What was she to do?"

And this was a heroic story - about how creative people are when, far from the home of their heart, they have to find ways to let the human spirit live even when all the props are missing! Here she was, in the middle of Brixton, with policemen at every street corner, and no sea for miles. So her African friends drove down to the sea at Brighton, on the south coast of England, and filled two big, plastic jerry-cans with salt sea water from the chilly English Channel and drove back to the metropolis. That night, they put a big tin bath in the local supermarket car-park, filled it with sea water and then walked their naked widow down to this substitute sea at three o'clock in the morning, closely surrounding her as she walked to shield her from passers-by or intrusive policemen, so that she could finish her purification in the ritually correct manner. Nimy concluded her story "I saw her a week later, and she looked wonderful: a new woman! A goddess! You could see that that ritual really had been a deep renewal for her."

In the old days, the stories people told at the wine harvest in these parts were not so exotic, but the *style* is the same as ever!

October
The rosary at the Chiesetta delle Ripe

There's a little side valley down in the woods where a stream ran years ago. Then the stream broke its banks, cascaded down the precipice and the valley was left high and dry, and the old mill further downstream as well. There's an old orchard there, with overgrown apple trees, plums, mulberries and vines, all abandoned years ago. When we still didn't know the place names, we called it "Paradise Valley" because it is so beautiful, but then we discovered that the local name is La Buca -"The Hole". I think this must be an instinctive effort to protect these places, rather like the old tradition amongst African mothers of giving their newborn babies ugly names at first so as to protect them against the attentions of Death.

At the end of The Hole, there's a little sanctuary with a fresco of the Madonna. The local farmers used to light a candle there, or leave flowers on the altar, or devotedly carve their initials in the fresco. From the grassy space in front of the sanctuary, you overlook the River Chiascio and the whole valley is surrounded by oak woods. The precipice of sheer rock is only a few paces away from the little church and the old people say that the sanctuary was built several generations back by one of Betto's family because the soil on his field was sliding slowly down the cliff, and when in desperation he prayed to the Madonna to stop this landslide, he had a vision of the Virgin standing at the edge of the rock. So Betto's ancestor built this tiny stone church to thank her for preserving his land: The Little Church of the Rock.

For many years, this little church was hidden in the woods and in the hearts of the people who still lived nearby, but a few years ago, Betto and his wife, who now live in town, decided to clean the place up and get the elderly priest, Don

Angelo, to come and say the rosary here once a month. At the beginning, we used to go too, with the children, because we liked the idea of all the valley people meeting in a wood and reciting prayers together in the great outdoors, especially since it is a prayer very like the ones addressed to the Great Goddess Isis. All the old peasants who had long since left the valley for urban life came back for this monthly gathering and after reciting the rosary, everyone had a glass of wine and ate the home-made cakes we all brought for the occasion. And that was a moment for listening to the old stories about the valley, told by people who had been born here. There was a strong feeling of community on those Sunday afternoons under the trees, although many of the men and women there now lived in the suburbs of Perugia.

I think our presence was mysterious to them. They knew we weren't Catholic and neither Martin nor I could ever find the words to explain our sense of the sacred nature of a community with its deep roots in *place*. They made good-natured attempts at converting us, but without knowing where to find our point of departure.

The stories we heard on those occasions were all about lateral thinking and creative living from past times. There was the story of Eusebia, the youngest of thirteen children, who owed her elegant name to a gentleman who happened to be passing their farmhouse just as her mother was desperately searching her mind for yet another name. When she grew up, Eusebia got engaged to a local boy, a good lad called Ubaldo, who thereupon left for the war against the Austrians. While he was away, Ubaldo's brother Gettulio up and married Eusebia. Time passed, and Ubaldo came home from the war but instead of getting angry, he simply joined the menage, and so they lived all three of them together, into a ripe old age.

Even more imaginative was Salvatore, Betto's long-gone grandfather. The old man who told me the story said Salvatore had the whole valley guessing for a year. "He told us all to keep the goose feathers for him! Said he'd come round every month and collect them, which he did, all over one summer, with a big sack. My wife would say to him, every time "What for do you want these goose feathers, Salvatore?" But he would just scoop them up and smile. "You'll see by and by", he'd say. All the women had these little mounds of feathers in the corner of their barns, waiting for Salvatore and weren't they *cross* that they couldn't find out what he was up to! "It can't be an eiderdown, he's not interested in the down, only in the proper feathers!" they said to each other. Even Salvatore's wife Igina was in the dark. "All I know is he's making an awful mess in my corn shed with that heap of fluff!" she'd say to the other women. When the winter came, that's a time when we can't do much, when it rains and snows, and then Salvatore locked himself up for hours at a time in the barn. "What's he up to, mama?" asked Salvatore's sons and Igina said, "Oh, don't ask me, he's doing something with those damn feathers - you know your dad, he gets *ideas*...as long as it keeps him happy!" People would make excuses to go past their place and peer through the window at him, but no-one could see what he was doing in there. Then, when it came spring, one day after Mass, he said "Who wants to see me *fly?*" We looked at each other and thought *"The feathers!"* He said for us to come to the rockface after lunch and there was a whole crowd of us waiting for him when he suddenly appeared on top of the cliff, with his arms stretched out - and he had a *huge great pair of wings!* That's what he'd been making all that time! No-one even had time to shout - he jumped!"

"What happened?" I asked, amazed.

"He broke both his legs," replied the old man, matter-of-factly. But then he added, with a look of awe and endless admiration, "But for a moment, up there against the sky, he looked like an *angel!*"

Icarus lived here too!

I turned round and, in the distance, I could see the evening sun glittering on the waters of the Chiascio river and I said to the old lady next to me "What a beautiful place this is!" and she replied "They're terrible places, but, there you go, we were born here!"

November
Washing by hand

"Mama, in London I keep hearing people moaning about having to do the washing, and I say to them, *"What* washing? All you've got to do is press a button on the washing machine!" They've never done washing in their lives!" Thus speaks our eldest daughter, who grew up here, where everything is done by hand, and now lives in London and works as a manager. She is constantly amazed to discover that the easy life seems just as hard to those who live it as life down on the farm.

It's true that we once had a washing machine. Martin invented it after an argument which ended with him declaring that *he* would do the washing in future...by the end of three days, we had a washing machine. It worked by pedalling an old bicycle. You pedalled for forty minutes and there was the clean washing! At first, like all novelties, it was a wonderful toy. At the first demonstration, there was even a large audience. It was summertime and there were lots of small children - ours and other people's. They sat excitedly on the courtyard floor in a row in front of the washing machine, as if it were a television screen, and enjoyed the show of coloured clothes swirling with the bubbles, appearing and disappearing in the little round window. "There's my skirt!" "Look at Ben's trousers!" It was a major source of entertainment for these children who had never had either a TV or a washing machine in the house. But after a while, the audiences fell off, especially when I began to mete out sessions of pedalling. "Ten minutes each and we'll have done it!" I would appeal. In the end the miraculous washing machine fell into disuse. If the choice is between pedalling nowhere for forty minutes or playing with bubbles, hands in hot water, I go for the latter as more fun.

No-one loves a pile of dirty clothes, and I don't say I love washing by hand. But I'd like to tell you how it feels, since I decided not to have a washing machine as a contribution to lowering Western standards.

In the summer, sitting astride an old bench with my sister with a tub of sun-heated water between us, we pass quiet hours under the mulberry tree, telling each other about our lives while we wash the clothes. We feel this is the height of luxury because only a few years ago, when we had not yet found the third spring down in the woods, we used to have to go down to the bottom of the woods, leading the donkey laden with dirty clothes, and wash them in the stream while the children splashed in the water, turning it steadily muddier as the day wore on. We have a vivid memory of the time Otello slipped the rope just as we were about to load him up with the clean washing, and Jane had to run all the way home to fetch him back to his duties.

But now, in the wintertime, with the wood stove burning and plenty of hot water, all I have to do is turn on a *tap,* pour in the liquid soap, turn on the radio and plunge my hands into the hot, soapy water. I look out of the window at the olive grove and easily fall into a trance. As if in a dream, the sheep and donkeys drift past my line of vision, and birds flit by, especially a robin who comes each winter and sits on the plum tree outside the kitchen, waiting for crumbs. When he has eaten, he always comes and sits just in front of me, on the window sill, as if to say thank you. The first year that he came, he threw himself with all his tiny weight against the window pane, as if to call our attention to the snow and his hunger, and when he comes back, he comes close to us, sits on the mulberry tree in the courtyard and sings. Like all quiet, repetitive activities, washing is meditative and calms the mind and the easy proximity of the robin seems a proof of this.

Over the years, I have realised that the washing has a double function in my life. Naturally the prime mover is the need for clean clothes. But gradually, as the pile of dirty clothes in the basket grows, a heap of disordered and chaotic ideas is growing in my head, thoughts that I have not had leisure to examine carefully yet. I read a great deal, and very diversely. I'm one of those people who reads six or seven books at a time and collects, obsessively and continually, good stories from friends and acquaintances. The three or four hours I spend washing clothes are a sort of appointment with myself, a meditation during which, without any particular mental effort, the connections appear spontaneously between the various ideas I've devoured since last time. I have come to believe that the rational activity of thought is not what we imagine it to be. We think of it as a sort of logical process under our conscious control, organised by a commander, the self. But when the mind is empty and we have a simple job to do, unexpected images and links float to the surface of the mind from unknown places.

As my hands deal with the first clothes, the least dirty ones, phrases of Plotinus' discourse on Beauty come back to me, episodes from the lives of the community in the Appenines called the Elves described in the thesis of a sociologist friend, horrible scenes of suffering in the early Australian penal colonies in the book by Robert Hughes, a poem by Kostantinos Kavafis called *Itaca* ("Above all, do not hasten the journey: let it last for many a long year, so that when you set foot on the island in old age, you will be rich in treasures gathered along the way. Do not ask riches of Ithaca. Ithaca gave you the wonderful journey, without her you would never have set sail..."), a silly detective story I read in the newspaper *La Stampa*, but not so silly after all if the images of the murderous lorry driver are still so vivid in my mind! And a text written by a friend entitled *A bioregional vision of the Region of Lazio* and the translation by another friend of the book by the American poet Gary Snyder, *The Practice of the Wild.*

I enjoy the smell of the Marsiglia soap, the feel of the hot water that I add as it cools, the pile of wrung out washing as it grows higher, the pile of dirty clothes as it disappears, the thoughts that come to me about members my family that their garments suggest to me and meanwhile I feel as if I am present at a relaxed debate between all those people whose words I have heard or read, each one with his or her light that illuminates the others. By the time I reach the end of the washing heap, after a long series of mud-embalmed socks, I feel as if I have not only a tub of clean clothes, but also a lucid and pristine mind.

You can't get the same effect by pushing a button.

December
Signorina, would you like a priest in your bed?

It's already dark when I drop in on Gina and I can feel the frost crunching on the grass under my feet. But in Gina's kitchen it's warm, the woodstove is on and there's a fire in the fireplace that her husband Orlando built in the corner just before he died. We drink coffee and Gina tells me she's going to kill the pig in a day or two and counts up all the relations who'll be coming over to help her process the meat. It's hard for her to manage alone now, at her age. She has to find neighbours who'll plough and harvest her fields, and people to help with the grape and olive harvests, but despite her poor knees which torment her, she soldiers on, keeping a vegetable garden going, looking after the pigs and chickens and hunting down the bastard porcupine who eats her potatoes. She eats him if she can catch him: "The last one weighed twenty kilos! No wonder, with all those potatoes of mine he stuffed himself with!" She laughs, because she knows she's as crafty as the porcupine.

While we chat, Gina fetches an old pan from her bedroom and fills it with embers from the fire. "I'm going to put this priest in my bed" she says "Maybe it'll ease these old knees". The wooden frame with the pan of embers they call "the priest" round here is a brilliant way of warming a bed, far more effective than a hot water bottle: when you get into bed the welcome is truly heart-warming. But every time I see one, it reminds me of the first time I met an Umbrian "priest", when I was still travelling round Umbria in the middle of winter, looking for a piece of land to buy. At that time, I was working in Rome and every weekend, I'd set off alone by train and get off at random locations in Umbria late on the Friday evening. Then I'd spend the whole weekend walking or cycling around the countryside, looking for a remote, small farm that didn't cost too much. I never took friends with me on these trips, because the dream was too fragile and too important then to share it with other people.

On one of these journeys, I fetched up in a village around ten at night. It was very cold and the streets were deserted. I found a bar next to the church and went in. There was a pleasant-looking, maternal, middle-aged woman behind the counter and I asked her if there was a pensione in the village. "We have a few rooms here, if you want one..." she replied. I was so relieved, because in some villages of this size there's nowhere to stay at all. She showed me up to a rustic little room with a big, high bed and then she said "Have you eaten? Shall I make you a plate of tagliatelle?" Oh, what a saint! She sat me down at a table in a back room and set before me a quarter of wine and a basket of that saltless bread they make here in Umbria, which I used to sprinkle with salt because I wasn't used to it then. How good I felt on those Friday evenings, at the start of another solitary adventure, with the peace of the night around me and unknown countryside to explore the next day. I enjoyed the freedom of eating alone and planning the next day with no-one else to think about!

The tagliatelle were delicious and I was half way through them when the woman came back and asked, with a kindly smile, "Signorina, I was thinking - would you like a priest in your bed?" I was stunned. A whole series of anthropological questions flashed through my mind. "Can it really be that these small villages have customs like this? Is this some sort of ecclesiastical *jus primae noctis*? Perhaps I've given this woman the wrong impression somehow? And yet she seems so respectable! And she asks me the question as if it were absolutely routine! Maybe she just wants to give her clients an all-round service? Maybe this is how they keep warm in winter in these rural parts? Maybe she's a friend of the Perpetua?". All these thoughts went through my head while the woman stood smiling and waiting for a reply. "Yes, please" I said, at last, because I couldn't believe she was really offering me a flesh-and-blood priest and anyway, it's always best to say YES to life. "After all" I said to myself, "I can always change my mind and say NO later, if I don't like the look of him!" When I got upstairs, I was delighted to discover what an Umbrian priest is, but at the same time a little disappointed that I hadn't stumbled across some bizarre, ancient, country habit!

After seven years of Roman life, the impact with country ways and a slower pace of life was full of surprises. I was used to city talk, speedy and to the point, but in the country I came up against different habits and discovered that conversations needed long pauses for reflection. Once I met a farmer engaged in spreading muck on a distant field. I got into conversation with him and after a while I asked if he knew anyone who wanted to sell a piece of land with an old house. "No." he said. We chatted for another twenty minutes, about the muck, the lie of the land, the weather, the state of the dirt track, my very old rented bicycle and about the old man down in Gubbio who had rented it to me. I realise now that this was the time he needed to have a good look at me. Suddenly, he said "My cousin down there wants to sell his land" "Really?" And five minutes later, "My brother wants to sell *his* land, too" And then, "Actually, I want to sell *mine*...why don't you come home, it's lunch-time, come and eat with us". I remember that huge old farmhouse, the wife and five children, all barefooted, and how we ate mostly- broth minestrone round the table, standing up because there were no chairs. The kitchen was vast and the wooden beams were dark from years of smoke, and there was nothing in the room except the table and the cauldron on the fire. It was 1975, and I could understand why this family longed to leave for the city while I longed to travel in the opposite direction.

When I first found this place, where I have put down roots, I was accompanied by the old couple of ex-peasants who had abandoned the farm a few years before. They drove me down the dirt track and into the valley in their old *Fiat 500* car which bounced over the huge holes in the road. When we got down to the farmhouse, I thought they would produce a key so that I could see inside. Even though the deeds to the land didn't even mention the house, except as "an old rustic ruin", I though it was part of looking at the place. But they had no keys! "How will we get in, then?" I asked. Francesco and Olga smiled in a relaxed and kindly way...they seemed to have all the time in the world. "Luigi's got the key" they explained. "Who's Luigi?" "He's the shepherd" "Ah, I see, but where is he?" I asked, looking round at the hills and woods where I could see no sign of any human being for miles. "We'll call him" Olga said and she and Francesco put their hands to their mouths and yelled, in sonorous voices used to calling to cows across distant pastures, "Luigi! LUIGIII!" Minutes went by and there was no sign of anyone. "Right, he's not here, we might as well go" I said, impatiently. "Oh, he'll be along in a minute," Olga reassured me. They explained that Luigi didn't live here, he just used the house to shelter from the rain or sun when he happened to be this way with the sheep. Well, there we stood in the middle of what looked like nowhere to me, but after a good ten minutes, a little dog appeared at the top of the field above us and ran down towards us. "There's Luigi's dog!" said Olga and Francesco triumphantly. The little brown and white dog stopped on the rock above our heads and looked down at us. Three minutes later a bigger dog came in sight. "There's Luigi's *other* dog!" But still no sign of the man himself. We looked at the dogs and the dogs, sitting side by side on their rock, gazed down at us.

Another quarter of an hour passed before Luigi appeared, but the old couple had relaxed now and they talked of other matters. Then we heard the ringing of sheep bells and a large flock of Sardinian sheep came slowly into view. They advanced down the hill and at last the silhouette of a man appeared, walking behind the flock.

I will never forget the first time I saw Luigi. He came unhurriedly down the hill and came to a halt behind the two dogs, on the rock. He was wearing an enormous overcoat which reached almost to his feet, and a cap which left the upper part of his face in shadow. From his shoulders and waist, there seemed to be a great diversity of things hanging: a woven wool bag, rope, a billhook. He stood there, motionless, silent, and watched us. "Luigi, we need the key" said Francesco. A long silence followed this request. And then, after several minutes had gone by, Luigi raised his chin and said, "*What* key?"
I was fascinated to see someone take their time like this. Finally, he reached slowly inside a deep pocket and drew forth an enormous, medieval key and threw it down to Francesco.

The signing of the deeds to the land was a real film. Where I worked in Rome, I was used to the smooth, urbane manners of the company lawyers, but I discovered that country lawyers are a different species, dealing with earthier affairs and at a more organic pace. They speak of "active and passive servitude", "ovine cattle"and haybarns that have been "used and enjoyed". On the day of the *compromesso,* which is the signing of an initial agreement to buy and sell on a given date, I came to the notary's office alone but the sellers were twenty strong, what with both near and distant relations, friends and henchmen of various sorts. There was Olga's ancient father, Giuseppe, who was the real owner of the land. He sat in a corner in a place of honour, with his stick in his hands. I thought we had gone over all the possible details but that afternoon I realised that selling land is not like selling an object. It is the passing on of a whole life, the passing on of an intimate relationship with a place, the whole history of a family. It is something so painful that the drama has to be expressed and acknowledged by everyone, but especially by the buyer because she is buying something priceless, something that can't be paid for with money, something that will in fact have to be deserved over the coming years.

A furious fight broke out between Giuseppe's two daughters. "You only keep father at your house because you want to have your hands on his pension!" "And you have always left him to me, you washed your hands of him!" they screamed at each other. Relations, friends and henchmen yelled even louder, in an attempt to calm things down. I felt very bad for the old man, but after half an hour of accusations, he rose to his feet and, shaking his walking stick over his head, he thundered, "I'm going to take this money and DRINK EVERY PENNY OF IT BEFORE I DIE!" Meanwhile, another Sardinian shepherd had appeared, Tonino, and I now discovered that he was Luigi's older brother and that the two brothers had a verbal contract for the pasture rights of the fields I was about the buy. I thought this hitch was something the sellers would have to sort out, but at that point, all faces turned in my direction. It was obviously up to me! "Maybe we could make a three-year agreement for the pasture and then see how it goes after that?" I suggested. "HUH!" replied Tonino sternly, "I'm signing NOTHING!" and he turned and strode out of the notary's office, slamming the heavy, carved door behind him. The Notary, a dignified, old-world man with an air of tolerant irony, turned to me. "Would you like to continue today, or would you prefer to come back next week?" he asked. The family was ready to call it a day and go home, but I had travelled up from Rome so many times to discuss things that I was determined to reach a conclusion. "No, either we sign today or that's it!" I said. "Well in that case we'll have to go down to Tonino and Luigi's house and talk it over" said Francesco, ever patient and smiling.

The twenty people in the party, all dressed in their Sunday best, got back in their cars and we drove back down into the valley, following the dirt track until we reached an old stone house on a rise overlooking the river Chiascio. The cars stopped, but no-one got out. "Why don't we get out?" I asked . "Well, you know, Tonino may be at the window with his shotgun" said Francesco. "Of course he isn't!" I said, and got out. Tonino, diffident but courteous, let us into the big kitchen and we sat down on the folding wooden chairs ranged around the walls. Luigi was out with the sheep. The others didn't want to wait for him to come home, but I said, "It really seems important that both brothers should be present" and so an embarrassing wait ensued. Olga made attempts at conversation "And how are the sheep doing, Signor Tonino?" she enquired. It seemed a bad subject to broach since the whole problem revolved around their pasturing.

Luigi came home an hour later. Once more, I waited for the others to start the discussion but again, everyone looked at me with interest, as if to say "Who knows what she'll say now?" and at that moment, I suddenly realised what the real problem was - it was a matter of respect. Tonino was the elder brother and he had not been consulted first. So I repeated my proposal more carefully. "Signor Tonino, I'm very sorry that there seems to have been a mistake here. You should have been consulted before. I was wondering if we could make an agreement for the pastures that would hold good for three years and then we will re-negotiate?" "Three years?" replied Tonino immediately, "That is fine by me!"

And so we got back into the cars and drove back to the notary, who was smiling patiently and drumming his fingers on the mahogany desk.

At the end of the day, I felt as if I had begun to weave my life into the lives of the moral owners of Pratale. All the bitter accusations were forgotten in the celebrations, I had proved myself ready and willing to take part in the game and the drama was over.

STAYING AT HOME

"Our age is retrospective. It builds the sepulchres of the fathers. It writes biographies, histories and criticism. The foregoing generations beheld God and nature face to face; we, through their eyes. Why should we not also enjoy an original relation to the universe? Why should not we have a poetry and philosophy of insight?... The sun shines today also".(R.W. Emerson, essay on Nature).

"The most radical thing you can do is stay at home" This sounds like an unbelievable piece of advice, coming as it did from the man who was mentor to the young poet Gary Lawless at the beginning of the seventies, Gary Snyder. Snyder was a key figure in a generation for whom the road, the trip, was paramount: a friend of Kerouac's and a man who had travelled widely with a knapsack on his back, worked as casual labour on tankers and up mountain tops as fire watch, a poet whose writings refer to *place* (*Mountains and Rivers without end* is long-meditated series of poems), Snyder had also undertaken a long journey to Japan and a parallel inner journey into Zen.

Lawless went home and committed himself to life in his native Maine, with the conscious intention of *listening to his place*. "Our bodies are still carrying on a conversation with the planet, but our heads are denying it." He wanted to find out what he would discover if he opened himself to his place, and his poetry communicates this conversation which he carries on, in a spirit of humility, with the sea breeze, white pine, the mists, the ice, the loons in the lakes of his home.

In *Looking for rivers*, he tells us the secret, what he found out. It is interesting to see that it is a perception that only poetry can explain - no work manual will tell us this - but it is the secret that could save our humanity. And it is this - *we are not alone here: there are what the American Indian peoples call 'all our relations', all the other living beings, and they do not ignore us, they speak to us - if we are listening.* "Birds speak as they fly overhead. Everything speaks. We live an uninterrupted conversation of life and light, light filtering through the stones, through the fog, gleaming off ice, warming fur, warming skin. ...Sometimes I don't know if I'm caribou or whale, man or woman..."

This may seem a romantic fantasy, but actually something real and important is being said here. It is an attitude of mind that has important practical consequences, as I found out for myself.

Ever since the time when I bought this abandoned farm near Gubbio in the foothills of the Italian Apennines - twenty hectares of pasture and woodland and a stone farmhouse dating from around 1600 which was only habitable in a very rough sense - the place has been slowly moving into my consciousness in ways I could not have foreseen at the beginning.

I left a job in Rome which had allowed me a good view of the multinational pharmaceutical world where I was beginning to see the devastating effects of globalization, and came up into these hills with my three-year-old daughter. At that time, I hadn't the slightest idea how to light a fire, which end to hold a hoe or the difference between a tomato plant and a thistle. I came because I had a feeling that right livelihood was not possible in a city, with a salary, and without any contact with matter, without being able to plunge my hands into the manifold substances of the earth. I felt as if there were a veil between me and the world which prevented me from being part of it and I was sick of that world with no inner limits or feeling of necessity, where everything is for sale.

I belonged to the post-war generation who grew up with enough to eat, able to study and travel (at least by hitching) and on my travels, it became apparent to me that there are still people who identify with their place, who are part of it, who accept that they are limited and defined by the nature of their place. I felt such an acute lack of this belonging that I was prepared to live in this cold and rather dangerous farmhouse and learn. I told myself that the skills of human survival can't be beyond a human being. If I think back now to the many times that the shepherd here showed me how to build a fire that wouldn't go out, I have to laugh. It took me some time to accept that you can't cheat the physical world: either there's enough dry kindling for the fire or there isn't: it is not like the advertising department of a pharmaceutical company, where they do black magic with words.

During the years of backpacking, I think a lot of people found that the only real trip is the one which has its counterpart in an inner journey. If the parallel adventure doesn't happen, the outer journey cannot illuminate us. When we go somewhere where there are still real inhabitants, we feel we are close to a truth, to a possible revelation. And yet tourist places communicate nothing but angst. What is it we are searching for along the pathways of the 'third' world? Or in the old city centres? I think it is the sense of lives honed by necessity and tradition, ways of living shaped by place, 'intact' cultures. *But we can create this culture for ourselves.* We can decide to be limited by what is here, we can be inhabitants instead of customers.

"But you can't turn the clock back!" A great many people have said this to me over the years while I hoed and milked sheep or washed clothes by hand. But I think we can *choose*: the Western lifestyle is not an inevitable road that the whole world has to travel, and it doesn't seem healthy or satisfying or *fun.* We can also look over our shoulders and ask ourselves if there were places along the way where we took wrong turnings. The revolution which changed us from nomadic hunter-gatherers to Neolithic agriculturists made our lives less precarious. The reappearance of the Paleolithic shepherd-hunters in the guise of kings (racketeers!) carried the Neolithic village towards the city and the state. The symbiotic association of humans, animals and plants in the Neolithic might be an acceptable level of civilisation. The culture that followed, the culture of the plough and food-surplus, brought with it the figure of the warrior king who exploited his own subjects: it was the beginning of the managerial revolution, made up of conflict and the domination of the few over the many, and this type of hierarchical organisation is still with us now.

It would be naive to see the Neolithic as the golden age! But we have to admit that there was a sort of 'living with little' at that stage which we might take as a model for future living, if we want to avoid the consequences of the globalization process in which a small group of capitalists are free to destroy the creative life of the majority, along with the other living species and the body of the earth itself. Lewis Mumford in his study of the evolution of the city says: "The decision which now confronts man (is)...whether he shall devote himself to the development of his own deepest humanity or whether he shall surrender himself to the now almost automatic forces he himself has set in motion. That second choice will bring with it a progressive loss of feeling, emotion, creative audacity and finally consciousness".
So this "living with little", hoes instead of tractors, hands in soap and water, visiting friends instead of using a mobile phone - is it so important?

It is what makes the spirit flourish! The gifts of poverty have been recognised all through the ages. Cabeza de Vaca, a Spanish gentleman and invader, shipwrecked in the 16th century in the Gulf of Mexico and washed up on the beach of the New World naked and unarmed, walked across the continent for 8 years. The first Indians who met him gave him to understand that either he cured their sick people or he was a dead man, and he, in desperation, made the sign of the cross over them *and healed them,* and continued thus for many years. When he returned to Spain, he wrote a description of his experience for the king and he concluded that the riches of European society, the luxury and comfortable living, the faculty of ordering others to do everything for us *takes away the individual's own power.* Only when a man is naked, hungry, alone, without help or defence, without anything, does power flow naturally through him. After eight years, Cabeza de Vaca, arriving on the Pacific coast, met a Spaniard at the head of a procession of Indian slaves carrying stolen treasures, and saw, to his horror, his own self as he had been eight years before. (The king of Spain, on reading these reflections of Cabeza de Vaca's, had him imprisoned until his death).

In the Hopi creation legend, the spirit-guide Màsaw tells his people to undertake a series of migrations in order to purify themselves before making permanent settlements - and advises them to choose a difficult place, with no water, where they would never forget the spirit. The clans carried out these migrations - actual, historically demonstrable journeys - some as far as 'Canada', Tierra del Fuego and to both American coasts, before settling in the vast arid plateau between the Colorado and Rio Grande rivers, where life is hard but where Indian religious culture holds its own perhaps more than anywhere else on the continent.

Francis of Assisi is a local example for us of the voluntary renunciation of luxury, and of the extraordinary vision of the rich spiritual life which flows from this gesture of limitation. "I will take the noblest and most beautiful spouse you have ever seen, whose beauty and wisdom will outshine all other women" - he was speaking of 'Madonna Poverty'.

It's always a good idea to pick role models well out of reach! My first teacher here in the country was Luigi, the Sardinian shepherd who pastured his sheep on my land (mine in the laughable sense that I had paid for it!) Gradually I realised that he lived in a very different world from mine. Here we are, both sitting eating his pecorino cheese on the same hillside, with Gubbio only a fifteen minute car journey away. Yet he went into town maybe once every four years. In those days it was an effort for me not to run into town every third day. He found no reason to go there - he didn't need anything.

This 'not needing' is something I have only partly learnt, and that very slowly. At the beginning, I was lucky not to have much money left over from buying the land. (A historical scholar from Gubbio, who once gave me a lift, pointed out that the reason for the intact medieval beauty of Gubbio is that until very recently nobody here had any money "If we'd had money, we'd have knocked down the thirteenth century Palace of Consuls and built ourselves some nice skyscrapers!") We are used to asking ourselves all the time what we need, what we ought to buy. *Not* to do this seems to us an actual sign of helplessness, of a disorganised life. My companion Martin, who escaped from an easy life in Zurich, points out constantly that we can ask ourselves instead, *"What can we do without?"* But it's a lengthy process of self re-education. What does it matter if our clothes are old, if our shoes are worn, if we walk instead of ride, if the nearest 'phone is 4 km away, if we eat what's here instead of what we can think up? By behaving like this, we begin to *see* this place, with its limits and its possibilities, we can see the *shape* of life here, which is also an aesthetic experience.

For example, there is a population of wild boars here - mammals as big as we are. They don't go to Gubbio, not even once every four years, and yet they are sleek, healthy animals. How do they live, what do they eat? It may seem very down to earth, but when we start bearing in mind the local food chains, the origins of the local culture become available to us, even though we come here from the supermarkets, pharmacies, and fast-food stores of the developed world with no basic culture of place. I heard a good story about a woman anthropologist who landed on a Polynesian island in the sixties. She learnt the basics of the language and started in with her questions, trying to find out from the women about their activities and beliefs. After several months of this grilling, the women decided to go and visit the Professor in her hut, and they were made very welcome. Full of admiration, the local women examined the Professor's possessions - drinking glasses, dresses made of materials unknown on the island, a clock. "How do you make this?" the women kept asking, since they were used to making all their baskets, cooking utensils, clothes and jewellery themselves. "I don't know - you see, we just buy them" their host kept repeating. By the end of their visit, the women had come to the conclusion that this Professor was the stupidest person in her own society and had been sent to ask them questions in order to learn how to do something useful.

That is how I felt. Our society is full of complex technology but we are mostly customers, not creators. Human culture emerges from the communion of hand and eye in the context of a specific place. The languages, songs, stories, art, and the outer forms of religious mysteries are human reworkings that emerge from a place, with its climate, vegetation, animals, foods, fuels, clothes, shelters. If we are not living materially in our place, we are cut off from the processes of human culture. The snacks we can buy are convenient but they are too fast and too alien to be bearers of spirit. Here there are still recipes for making acorns edible: they are slow and laboursome but it is food with multiple layers of meaning.

One of the presences with whom we live here is oak. Here oak is a god: huge, long-lived, beautiful and strong, the welcoming sustainer of many animal and plant lives (fungi, mistletoe, lichens, hundreds of insects, mice, pheasants, green wood-peckers, jays, black squirrels, wild boars). The intimate living together with the oak gave rise here to a divinized oak in the life of the ancient Osco-Umbrians who inhabited these hills before the Roman domination: the God Grabo. This is what happens when people stay at home. They listen to their relations and come to express recognition of their powerful presence in human life. Thus instead of using the human mind to isolate themselves, they begin to participate more consciously.

There is a translation of an Inuit story by Edward Field which describes this sense of participation. The times they refer to are not necessarily in the past - it can be today.

In the very earliest times
when both people and animals lived on earth
a person could be an animal if he wanted to
and an animal could become a human being
Sometimes they were people
and sometimes animals
and there was no difference
All spoke the same language
That was the time when words were like magic
The human mind
had mysterious powers
A word spoken by chance
might have strange consequences
It would suddenly come ALIVE
and what people wanted to happen could happen

all you had to do was SAY IT
Nobody could explain this:
That's the way it was.

How unbearably sad that phrase sounds, 'when both people and animals lived on earth', as if we humans had moved far away, hardened our hearts, disappeared from the world where everyone else lives. But where do we want to live if not on this earth? Where we can feel the sun, the rain, the wind and the sounds that everyone else feels and hears? If we still lived in the company of the non-human others, maybe we could create a human culture which would fit in.

One day in mid-summer I set off to collect agrimony flowers in the fields during the hot hours of the morning. I walked for many hours and I not only lost my sense of time but also my own point of view. I realised that by concentrating on the agrimony, on its favourite sites, on the stage of flowering, on the way it had spread across the hills over time, I was seeing the landscape from the agrimony's viewpoint instead of my own. I am not using a metaphor; it was a very strange experience. I thought of the women who were persecuted as witches: if this sense of participation could arise after four or five hours of gathering, what could happen to a human being over many years of solitary gathering? So this is what I found out by staying at home: how to behave in the knowledge, with the *feeling,* that my human eye is not the only one in this place, how to live this place also as a crow, a freshwater crab, a juniper. This 'collective eye' is not only an inner experience, it has practical effects in the everyday farm work, it changes the relationship with the sheep, the insects, the plants.

Over the years that we have been here, my companion, passing friends and I have carried out a slow piece of work to transform the old cleared fields into what the Romans called a *saltus*, a wooded pasture. We have tried to make space for our needs (pasture, vegetable gardens, orchards, vineyard, olive groves) using this wider viewpoint, leaving wild areas and corridors, waiting for the trees that have begun to grow across the fields to be big enough to withstand the grazing sheep and horses, before we cleared the undergrowth around them. So each winter we cut the Spanish broom, wild rose, dogwood, pyrocanthus, blackberry, old man's beard and thistles (the natural colonisers of the cleared fields which precede the regrowth of the woods) around some of the bigger trees: common oak and turkey oak, juniper, elm, hornbeam, field maple, ash, wild service tree, sloe, false acacia, hawthorn, cornel cherry and more domestic trees such as cherry and maraschino cherry, olive, walnut, hazelnut, laburnum, apple, fig, plum, elder. These trees, scattered across the steep pastures, prevent erosion and offer shade for the domestic animals and 'habitat' for the wild ones. The leaves enrich the soil and the animals find leaves, berries, fruit and acorns to eat.

Usually in the country, the sense of 'beauty' is associated with the peasant tradition of 'cleared' land. *Pulito!* - Clean! said my tree-devouring neighbour Pepe with great relief when he had finished pulling out all the field maples in his old vineyard. The pressures of share-cropping have contributed to this attitude and what looks 'good' to the modern farmer is a landscape in which there is nothing thorny, tangled, rotten or unplanned. It was hard to learn to live with brambles and see them as something beautiful, but gradually we began to see the wild parts as a fertile chaos, a source of health: in that wilderness good things are happening.

Our work is beginning to show results, but it is a job that has a slow pace and only happens in restricted areas, so as to welcome uncontrolled growth of these colonisers (another Umbrian God, Poprs, derives his name from the word for growth). These plants are a protection for the soil and are important presences in themselves. We don't want to have everything under control, and the area that we have cleared will be wild again when we come back to it in many years time.

Lawless has a dream meeting (*Ice Tattoo* from *'First Sight of Land)* with a caribou, a conversation:
"We talked with him, wishing
meat for our families.
He warned us that our troubles came
from killing too freely .
We must kill only for food,
to renew tools, tents, clothing.
We must not waste life. Their blood is his blood,
their blood is our blood.
Their dreams are our dreams and he
is one of us."

I think a lot of people are discovering the delicate balance we can live inside once we abandon the romantic idea of human supremacy and embrace the more realistic collective viewpoint. The ability to feel the reality and importance of the other forms of life as if they really were part of us is the only way to learn to live without taking more than necessary, more than our share. It is a revolution. This communion with all our relations may lead us back to the place where we took a wrong turn and left the path for the motorway.
"Here by the road where people are carried, with
or against their will, as on a river of burning oil
through a time already half consumed, how
shall we pray to escape the catastrophe
that we have not had the vision to oppose and have
therefore deserved, and that many have desired?

Yet here in our moment in the ages of ages
amid the icons of fire from the maddened center
whirling out, we pray to be delivered from the blaze
that we have earned, that many desire. We pray
that the continent of love may be shaped within
the continent of power, here by the river of fire.

We pray for vision, though we die, to see
in our small imperfect love the Love
of ages, whose green tree yet stands amid the flames. May
we
be as a song sung within the tree, though beside us
the river of oil flows, burning, and the sky is filled
with the whine of desire to burn and be burned in the fire".

Wendell Berry,
Sabbaths 1989

January
Four Old Men

This morning, we walked up the snowy road to say Happy Birthday to Antonio. It's St. Anthony's day, patron saint of animals, and Antonio is eighty-seven. We found him sitting in his usual corner, by the fireplace, where a large oak log burnt slowly. "Is it St. Anthony's? Happy Birthday, Auguri!" we said as we went into the kitchen and Antonio laughed and said "St. Anthony with his long, white beard/brings the snowflakes down to earth!"

When we first came to the valley, Antonio was not the only Coccia brother, as he is now - at that time, the neighbourhood was made up of four brothers, all getting on in years: Antonio, Alessandro, Orlando and Artemio. Now Antonio is the only one left to tell the old stories, but they were all memorable people.

Today, while Rosa made us a pot of coffee, Antonio relived his younger days by telling us *how he stole Rosa from her fiancé*. That was a brilliant operation and it warms his heart now, in the weakness of old age, to remember that moment of triumph. "In those days, there was still the tax..." ("What tax?" we asked.) "The Fascist tax that all bachelors had to pay. Once you turned twenty-four, if you didn't marry, you had to pay up! There were lads who married just because of that tax. I recall Gino saying that if he had to pay the tax, he might as well set up house and get married! But that wasn't why I did it - I really fancied Rosa, she was a beauty when she was young! I was in the army then and I came home on leave, but Rosa was engaged to someone from round here. It only took me a matter of days to change her mind for her! But I left the straw, you know, as a sign. ("What straw? What sign?" we asked.) "That's what we used to do in these parts: if you stole someone's girl, you left a little trail of straw leading from the fiancé's front door to your own house! To show him who'd got the girl now! Hah! When he woke up in the morning and saw that straw...! Rosa was sixteen then, she was the prettiest girl in the valley. We got married straight away and I was off again, back to the army!"

"Yes, and here I am" grumbled Rosa, as she passed us the little cups of black coffee, "still being ordered about by Antonio sixty years later!" But she smiled, because she still holds that romantic beginning in her heart.

Orlando had been a prisoner of war in Scotland and he often put lost backpackers on the right path for our house. On these occasions, he would pour them a glass of his wine and amaze them with his explanations in English and his strong Glaswegian accent. He had spent most of the war in a Scottish warehouse full of food rations for the troops, so he and his companions had eaten well, cooking up spaghetti and drinking whiskey, pleased to be out of the fray. He told us how, when they were first caught, the British major asked them "Who wants to collaborate?" - "Well, of course, we *all* put our hands up!" said Orlando, to whom any sense of military honour would have seemed absurd. He was a peace-loving man and when his neighbours argued, would take it upon himself to go with a gift of his honey and try to make peace between the parties. He was the first person I had ever met who took onto his shoulders this sort of personal responsibility for a whole community. Once I met him walking down the road on one of these peace-keeping missions and he had a bee buzzing near his ear. "Orlando, be careful, there's a bee!" I said and he replied "No, the bees all know me, they wouldn't sting me." He was at peace with the world!

With Orlando, one could sit in a companionable silence. Once, peeling the leaves off the cobs of maize in his barn, I was sitting with him in an easy silence of this sort when a friend of mine from Rome came roaring up on a motorbike. He dismounted and came over to where we sere sitting and talked - fast - for maybe twenty minutes and then jumped back on his motorbike and sped off towards my house. Orlando and I continued to work in silence for another ten minutes or so, with only the rustle of maize leaves, and then Orlando said slowly, smiling, "How he does *talk*, that friend of yours!"

Yet Orlando himself had a huge repertoire of stories and ballads, and if one could persuade his wife to turn off the TV, and let Orlando get telling, it was fascinating to listen to him. I remember a long dialogue in verses between a Florentine urbanite and a Tuscan peasant in which the merits of the rural and urban lifestyles were discussed. The Florentine, who objected to eating in a tavern with the peasant because of the "smell of dung" that hung about the latter, summed up his point of view by saying that, if he were in charge, he would have the city gates shut to keep all villains out of the town, whereupon the peasant reminded him that in that case, all the flagons of good Tuscan wine, the forms of cheese and legs of cured ham would also be shut out. (I once begged Orlando to write this long ballad down for me and he spent a winter writing it out, slowly, in an old exercise book in the evenings.) Often, when Orlando recited one of these old poems, his brother Alessandro would act as the chorus, repeating at intervals, "This is all true history, you know!"

Alessandro was a wiry, energetic little man whom we often used to meet walking round the valley. He walked miles between his farm and the farms of his various brothers. The first time I saw him, he was driving an enormous white cow down the road. The cow was so much bigger than Alessandro, it made him look like a gnome, but the cow obeyed him because he was good with animals, and so she didn't seem to mind being pushed and prodded from below.

Alessandro had no wife, but he had an elderly lover who lived next door and he also shared a wine-cellar with this woman's husband, a happy arrangement that seemed to have lasted a long time. The lover was also the caretaker of the little church, and no one seemed to find any contradiction in this. The only comment I ever heard about their relationship was someone saying, rather enviously, "They're always stuck together!" Alessandro's wife had asked for a divorce many years before and her chief complaint had been that he never washed. The judge had come down in Alessandro's favour and there was a famous song which Alessandro had composed about the hearing, in which the judge praised the defendant's honest sweat. Alessandro would sing this song at festas sometimes.

But the thing that really touched my heart about this tough old walker was the way he acted as Father Christmas. Every year, just before Christmas, he would appear at dusk, in his long overcoat that almost reached the ground, and out of the deep pockets, he would produce little, home-made twists of paper full of sweets for all the children. He walked from house to house all over the valley, and never forgot any of the children. I was grateful to him for including us 'foreigners' in his open-handedness and he seemed to me to embody the very spirit of Christmas. He was a much realer Father Christmas than the ones you see in the supermarket, despite all their red and white costumes and beards.

During one of those maize-peeling sessions, Alessandro once began a story of his own about some local people, dead but not so long before. The formula with which he began the tale struck me so forcibly that I never caught the rest of the story. He said, "Once upon a time there lived two brothers. One was honest and hard-working *but the other was of sad grace...*" I thought to myself, "He's talking about real people who lived here not so long ago, people everyone knew...and yet they have already become part of the fairytale world!"

The last brother, Artemio, was a soft-hearted, kindly man, very religious. People said that during the war, he had escaped from the Germans by throwing himself out of a moving train, and that, as he lay between the tracks, another train had thundered over him. He got out alive, but in those terrible minutes, he had dedicated his life to the Virgin Mary who saved him. For many years, he used to set out from home, barefoot, every Friday evening and walk to Gubbio, twelve kilometres away, pray to the Madonna all night and come home at dawn to start the day's work.

But his Christianity was very practical. The first year that I had a vineyard, borrowed from a neighbour, Artemio came and pruned it on a Sunday morning, instead of going to Mass, because he was too busy during the weekdays and he knew I needed help. He also offered to come and plough the old vegetable garden, which hadn't been dug in eight years. I was still living in Rome at that time, but my mother and father were here on holiday, bravely roughing it in the old house. Artemio turned up one morning with two oxen and the plough. While he was getting the plough hitched up, one of the oxen raised its head and ate some of the grapevines that grew over the arched doorway. My father spoke no Italian, but he reported the scene to me. "Artemio stood in front of the ox with a severe look on his face and spoke to the animal for a good ten minutes, gesticulating and repeating the same words several times. I couldn't understand the words, but it was clear that he was telling the ox that they had come to do someone a favour and it was not right to snatch someone else's grapevine, that it was time to get working and behave himself. He spoke to him patiently and firmly, as if he were a person - a disobedient child .He reminded me of Pythagoras talking to the ox at Taranto!"

As he grew older, Artemio became more and more devout. He used to walk down to the little sanctuary at the Chiesetta delle Ripe and spend hours standing with his arms outstretched, his lips moving, speaking with the Madonna in the fresco. This maternal figure seemed to grow more luminous in his imagination until nothing else filled his mind, a situation which created endless problems for his family. But I think he left this life in a blazing state of inner grace.

These were the four old men who lived here when I first came to this valley. They sowed their gifts here for anyone who wants to gather them.

February
Imbolc

The atmosphere in the sheepshed is so peaceful at night. I like being there for a while after I have distributed the hay. I stand quietly in a corner and watch, as each sheep finds its habitual place at the manger and the lambs gather together and sit in a great heap in the middle of the stable. There have been a lot of twins this year, almost one twin birth in every two births and there are several speckled lambs. This is because of the ram, who is white but who had a black mother. So there are white lambs with black faces and legs, and black ones with white feet and tails and tufts on their foreheads. If you ask any questions about animals of the local people here, they always reply "You have to watch them and see" and over the years, I have come to realise that they are right: you have to have the patience to spend a lot of time just observing. It's the only way to notice who is limping because of a stone in the hoof, who's off their food, who's close to giving birth.

This evening, I made up a bottle of dried milk for a couple of newborn lambs. Their mother is old and I doubt she had enough milk for two. While I feed one of them, my friend Francesca holds his brother. Then, right at our feet, Rosy has a lamb. Francesca is delighted - she's never seen a lamb born. And the lamb is extraordinarily alive - as soon as it touches the ground, it gets to its feet and wanders off, as if being born was nothing to shout about. While we watch this little explorer in amazement, Rosy turns round twice and drops another lamb. They are both speckled, white, black and brown, both females. The second lamb is much slower than her sister, and it takes her a few minutes to notice she's in the world.

The sound of a sheep calling her newly born lamb is so throaty and gentle, it is as if she lowered her voice so as not to hurt their delicate little ears. It sounds - can one say this? - full of amazed love.

In the whole Celtic area, before the arrival of Christianity, there were four great festivals in the year and one of them - Imbolc -fell on 1st February. In Gaelic this means "in the belly" and this feast centred on the celebration of the lactation of ewes. Perhaps this word also referred to the coming spring, still hidden in the dark belly of winter. Every year, when we live through this dangerous and miraculous, birth-giving time, I feel how much the arrival of lambs must have meant to a pastoral society, and how important it must have been to celebrate this first sign of the fertility of the coming year. It is snowing outside but inside the shed there is a warm sense of palpitating life, a tangle of little woolly bodies jumping over each other and nestling up close. Out of the twenty lambs born so far, only one has died, and its mother is still calling for it with an anguished cry. Each time she bleats, nineteen little lambs reply, and she listens intently, her head quite still, but she recognises straight away that none of those bleats is from her own lost lamb and so she continues to push her way through the flock, backwards and forwards, searching for it in all the corners of the stable.

The Mongolians, who are a pastoral people, have a traditional form of song called Xöömii. It is a technique which originated in the Altai mountains: the singer uses his mouth as a soundbox, so that each fundamental note he emits is accompanied by its harmonics, which seem to the listener to hover in the upper air. One of these singers, Batzengel says "We imitate the murmur of streams, the mountain echoes, the voices of animals, thus we express the beauty of the world." These songs spring from a shamanic tradition and these sounds redolent with birdsong, galloping hooves and the bleating of sheep are also a bridge between our world and the supernatural world, a symbolic journey

undertaken for healing, communicating with the dead and for divination. And in fact the Mongolians say that they also use these sounds to heal sick sheep. It is interesting to see that this technique has arisen independently in different parts of the world (in Mongolia, Tuva, South Africa) and always amongst people who are herders, as if the animals had communicated something of their essence to the human beings who live along side them.
Listen to the sound of a flock of sheep and then listen to someone singing harmonics!

The first time I heard Mongolian overtone chanting, I was struck by its resemblance to Sardinian singing and I went and listened once more to the *Tenores de Bitti* singing.. Yes! There are the same sounds again, and the same body language too, the men standing close in a tight circle facing each other, their hands to their mouths, listening to themselves, not performing but intoning the songs as if they were a rite for entering the natural world, where men live close to the animals, as they have done for millennia in Sardinia and where many still do.

Many modes have arisen, over the centuries, for us humans to express our acknowledgement of the non human presences in the world: rituals and songs performed for the Gods of mountains and rivers, forests and seas, half-human half-animal figures who have danced in and out of our consciousness, a celebration of the lives of the other species we depend on. In our part of the world all this has disappeared now. As T.S. Eliot says, we have forgotten the psychological trick which allowed us to see visions. The air around us, that was once thick with spirits, is now empty.

In the morning, I go back down to the shed and find that Lilly has had twins, but in the night, one of them has got lost in the crowd and now Lilly won't have him, absolutely refuses to recognise him. He is cold, still wet and hungry, and he cries and cries. I rub him dry and then I hold onto the sheep whose lamb had died and let him drink from her full udder. Over the next few days, with great exercise of patience on our part, we persuade this wicked step-mother to soften her heart and accept him. And he survives.

For all these dramas of life and death, I wish I too had songs and rituals of participation. I wish I had other celebrants for this Imbolc, but I content myself with a corner of the stable where I feel part of the quiet evening.

March
Dyeing with lichens

It is quite extraordinary how Nature drags me into more and more complicated bits of work, although I already have plenty to do. It is as if, as soon as I had five minutes to stand and stare, Nature thought to herself "Here's a two-handed being in a horrible state of idleness...I'll give her something to stare at."

I was sawing up logs when Martin arrived at the wood-stack with the horse and unloaded a pile of branches from a half-dead tree which he had cut in the old vineyard. In Umbria, the traditional way of growing vines was to train them up field maples, a tree which the locals call "albero" which just means "tree", because of this function. We have hundreds of these field maples where there used to be an old vineyard, but the trees have grown too big and the vines have gone wild. We planted a new vineyard further up the south slope, but these old trees make welcome shade on the pastures and give us a few dry branches for the fire every year.

I picked one of the branches up and started sawing it, but then I was struck by the multitude of lichens that covered it. They were all shapes and colours: grey, yellow, orange, dark green, pale green and pale blue. They were beautiful, and as I looked at them, I remembered that up north, in Scotland and Finland, these lichens have always been used to dye cloth. They have the great virtue of being substantive dyes - they don't need a fixative as most plant dyes do: their colours stay fast in the wool without any alum or iron.

I put the saw down with the feeling that I had set my foot on a long path and went to fetch the book on lichens.

I found the book, but it was a very old one, and all the photos were in black and white. I would like to know who can identify a lichen without the help of a colour key! Their structures are so tiny and they all look alike. So I thought, "Never mind, I don't have to know all their Latin names, I'll proceed empirically, separating the lichens by colour and dyeing samples of wool and see what results I get."
In the course of looking up a recipe for dyeing, I came across one from medieval Scotland, for obtaining the red tartan for making men's kilts. It said to scrape a certain quantity of lichen off the rock, steep it in *old* urine with a bit of powdered chalk and then leave the brew in a warm place for *two months* (perhaps in the corner of the hearth?) In case you think this dye was the invention of Barbarians, let me tell you that it was a certain Signor Ruccelli from Florence who learnt this procedure in Asia Minor and passed it on to the Northerners! I decided I could do without flame red, but if anyone wants to try it, the lichen you need is Ochrolechia tartarea.

The simplest recipe said to put the lichens in a large pan in layers with the clean wool and fill the pan with cold water. You then put the pan on the fire and bring it very slowly to the boil and leave it to simmer quietly for several hours until the desired shade is obtained. Luckily at this time of year, we have the wood-stove on anyway to heat the house.

I cut the branches into short lengths and left them in a tub on one side. I had to defend them all day against people who wanted to burn them in the wood-stove. Then our neighbour's little girl, Giorgia, came down and the two of us spent the evening scraping the lichens off into six different plates, according to colour. The members of my family watched us with the air of people who have seen all sorts of crazy manias, so I was lucky to have Giorgia with her ten-year-old enthusiasm!

Over the next few days, I washed the wool from the last sheep-shearing. I had kept seven fleeces, in case I had time to use them. It is a strange experience, washing wool. I have to do it in the winter because we don't have any spare water in the summer, and washing raw wool takes many rinsings because you can't use soap, or you lose the lanolin which makes the wool water-resistant. When I put the wool in the water and plunged my hands in, it felt like liquid silk! It must have been the lanolin. When the wool was white, I laid it out on the wall of the courtyard to dry and then the experiment began.

Well, it was one surprise after another! The yellow-orange lichens gave a deep rose colour, the pale green one gave a deep gold, the dark green ones gave a pale peach colour, and a mixture of all the lichens gave a sort of brownish-beige.

So Giorgia and I felt pleased and we strewed the courtyard wall with coloured wool in heaps to dry. I had to keep threatening the cats and dogs, who would have liked to lie on these soft piles. The marmalade cat Red Jeffrey was particularly taken with the colour which matched his fur, perhaps thinking I wouldn't notice him thus camouflaged and would let him lie on that nice bed in the sun in peace.

It was only when the wool dried that I realised that I would now have to learn to spin! Otherwise why had I dyed all that wool? Well, so as not to waste the lichens, but had I got time to spin? Years ago, I had asked my neighbour to show me how to spin and she had taught me, although she was unable to slow her movements down after fifty years. I had spun a little, but at that time I was busy with home-schooling the children as well as the farmwork and I also wanted to have time for reading. I had decided than that spinning was a slow activity, from a pre-literate world, and I'd given it up. Now I got the spindle out and tried again, sitting in front of the fire like a crone from the olden days. I was pretty clumsy. My son Ben watched me with a sceptical air and then said, helpfully, “I haven’t got time to do that, but if I tried, in five minutes I’d be spinning like an old witch who’s been doing it all her life!” Well, thanks Ben!

Then I went on a long train journey to Bolzano and I took some carded wool and the spindle with me. My hope was that some old woman on the train would take pity on me and say “Oh, give me that spindle and I’ll show you!” But those old women don’t seem to be around any more, at least not on the Italian State Railways. The other passengers, horrified, watched me out of the corners of their eyes while I span and they recited monologues on their mobile phones. I was getting much better at it, the thread was coming quite smooth by now. I hoped no-one would complain about the strands of wool which flew around me and I thought to myself “If they say anything, I will complain about the dangerous waves from their mobiles!” But they were perfectly pleasant and by the time I had done the journey both ways, I had several balls of spun wool ready.

Yes. Ready for what? Oh, my God, I’ll have to learn to knit now. I went down to my bar in Gubbio and had a lesson from the woman behind the counter. I had seen her knitting beautiful jumpers between making cups of cappuccino, so with her advice and the help of a weaver friend, I learned the elements of knitting.

So now in the evenings, we sit in front of the fire and Martin plays the flute while I knit. I hope Nature is happy! The countryside, I now see, is full to the brim with lichens - I had never really noticed them before, and knowing their *hidden* colours has added a whole other dimension to my deep map of this place.

April
The Carpenter Bee

Today I built the structure for the little green beans to climb up: a line fifteen metres long of ash poles and juniper poles sunk into the earth and lashed together with string. While I was working, one of my neighbours came to watch, just like he did last year. He's a carpenter bee, *Xylocopa violacea*, big and black with a blue shimmer in his wings. This bee visits the black and white broad bean flowers and the borage flowers in the spring garden and then he spends all summer around the green bean flowers. Last year was the first time I noticed that he remembers the bean poles: as soon as I started putting them up, he flew slowly along the whole structure, carefully examining all the poles. I thought he was remembering the flowers and checking to see if his favourite plants had started growing but later I realised he is a *carpenter* bee, who builds his nest in wood, and he was really checking to see if these poles were suitable nesting places. Here he is again, checking them out: he obviously likes to work alongside someone else.

It is hard for us to believe that the other beings are interested in our activities and have means of communicating with us, but last summer when it came to the season for picking beans, I finally came to the conclusion that this carpenter bee likes company. While I wandered up and down the lines of beans, filling my basket, this bee would hover over the flowers nearby. At first I thought there was more than one insect but then I saw that when I moved to another line, the same bee followed, as if he were aware that we were both engaged on a similar gathering task and preferred to work in pairs.

Years ago, a friend said to me "When you talk to animals, remember that *they understand*." The best example I ever saw of this was when our donkey started chasing his own daughter who was on heat for the first time. I wondered whether his real "wife", Titania, would be jealous when she saw this hot pursuit, but then I thought "No, of course she won't, she's just an animal". During the morning, I happened to walk through the animal's courtyard on my way to fetch a hoe, and there was Titania, standing there alone with her head hanging down and her ears drooping, the very picture of misery. As I came close, I said in a cheery voice "What's your husband up to, Titania?" In a flash Titania leapt round and kicked me on the bum. I was shocked! She was always so affectionate and quiet: in twenty years that's still the only time she has ever kicked anyone! I was sobered and ashamed: how could I have been so thoughtless as to think she had no feelings?

In 1763 an American Quaker, John Woolman of New Jersey, undertook a difficult journey westwards to Wyalusing in order to meet a chief of the Delaware Indians, Papunehang, and talk to him. He gave as his reason for wishing to visit the Delaware Indians "I wanted to feel and understand their life, and the spirit they live in, if haply I might receive some instruction from them". When the meeting took place without an interpreter, Papunehang motioned to Woolman to talk English, even though he knew nothing of that language. The next day, when the interpreter was present, Papunehang explained: "I love to feel the place where words come from".

That place where words come from, that inner desire to embrace other beings, exists in the non-human world too and if you live alongside plants and animals, you can take part in the continuous conversation that flows between us all. You can take part in all sorts of surprising, non-verbal ways: Martin, as he walks across the pasture, can smell the slightly acid perfume of a horse on heat. It means knowing what is going on in the world.

"If there were no more bears, we would surely die from a great solitude of spirit" said Chief Seattle, who knew about that conversation and lived inside it.

The carpenter bee has reached the end of the bean beds I am working on and is hovering over the blue borage flowers and I stand and admire how the blue shimmering of his wings resonates with the colour of those petals: *everything* in this world calls out to each other!

May
Country births

“He’s born!” shouted our neighbour Ivo, smiling from his upstairs window. We were on our way home from the mortuary where we had said goodbye to a friend as he lay in his coffin, his face tucked round with delicate, white lace that had reminded me of a newborn baby in his cradle. This is a day of hellos and goodbyes that touch the heart: Leonardo, the most recent member of the valley community, has arrived and there’s a blue rosette on the house where Lucia gave birth eleven times. Only this time the
new inhabitant has more distant roots because his mama Masako is from Japan. He is a sweet baby and he reminds me of other country births.

He reminds me of the birth of Beniamino, a real group event. The local hospital is small and quite friendly but it is hard to give birth there without interference so I had thought out a system for foiling intervention: when I went into labour, I went down to Gubbio with a large crowd of friends and we sat and sipped drinks and chatted in the bar opposite the hospital. Every so often, I dropped out of the conversation to float with the contractions into another dimension and as the morning advanced, I could see the barman looking more and more anxious as he came out with another tray of drinks.

Back home, we had a large pan of pear jam on the stove. My sister Jane and I had decided we needed to add serious amounts of lemon juice because pears don’t have much pectin and the jam wasn’t setting. So while I chatted and breathed through contractions, Jane went over the piazza to the market and bought a huge plastic bagful of lemons. By midday, the baby seemed on its way at last, so Jane and I and my friend Suzi said goodbye to the Labour Committee and staggered into the hospital. The midwife was a wonderful woman called Archangel whose presence was so commanding that all the medics kept right out of her way and I didn't have to defend myself as I had feared. However, once Ben was born, they gave us a room to ourselves so we wouldn't contaminate the other women with our wild ideas on childbirth. It was a lovely place to sleep, overlooking the warm summer night on the main piazza, with the street lights shining through the trees, as if we were in a first class hotel! Next morning the head doctor looked round the door and cleared his throat "Ahem, signora, do you mind if I ask you a question? *What were the lemons for?"* "The lemons?" we replied, amazed, "They are for the pear jam". He obviously suspected that citrus fruits were in the nature of a secret ingredient for natural childbirth. Then Ben's big sister Melissa arrived with the gang and we set out for home with the baby and the bag of lemons. That evening, all the neighbours drove down on tractors, bearing flagons of red wine, and celebrated Ben's arrival until the small hours, while Luigi waited patiently, sitting on the field in the dark, to see his son for the first time in peace and quiet.

Some years later, Jane gave birth in the same hospital and we always remember the dialogue as the baby was about to arrive. "Etain, it's so hot, can you open a window?" gasped Jane and I went to open it. "WHAT are you doing?" asked the doctor and midwife, amazed. "It's so hard to breath with this heat" I replied. "But this is a hospital, you *can't* open the windows!" they said sternly. I saw that what they said was true enough: the windows were screwed shut. So Jane, already *persona non grata* because she insisted on squatting instead of lying down on her back, took off the big cotton Tshirt she was wearing. The medical team recoiled in horror. "What is she going to do, give birth NAKED?" they asked me. "Yes, what do people here usually do?" I asked. Then little, half-Caribbean Solomon was born and everyone forgot about the outrageous nakedness because he was beautiful and everyone wanted to cuddle him and he was happy spread-eagled on Jane's bare tummy. Later that afternoon, I went to buy Jane something to eat, and when I went by the pharmacy, the woman behind the counter whispered to me "Was it your sister who gave birth *with nothing on?"* The whole town seemed to have heard that the wild barbarian women were amongst them.

It was to avoid all this that our friend Monika from the next valley decided to have her baby at home on the farm with no-one but her partner to share the experience. A birth isn't just one day in our life, it's a part of our selves, something held forever in the heart and it is a terrible thing to let someone steal it away. Our own farm animals fare better in this than we women do. So when the time came, Monika walked up and down her garden with Alan as the labour pains grew stronger and stronger. She told us that there was a moment of great confusion when it seemed that the baby would never come out. "Then I went into the house and suddenly I knew exactly what to do. I got down on my hands and knees and the baby slid out like a little salami. It was so wonderful, afterwards, to be at home together, all of us, in the shadows at dusk, cuddling Nico in the quiet!"

Alan and Monika told me two stories about Nico's birth.

A couple of days after he was born, Alan went down to the Registry Office in town to declare Nico's birth. "Right" said the clerk, his pen poised over triplicate forms, "So: name, surname, your date and place of birth, your wife's date and place of birth?" Alan dictated these awkward, foreign details to him. "Midwife's certificate?" asked the clerk., "We haven't got one" replied Alan. "Oh no, you aren't following me...we'll start again. Name, surname, birthdates *and midwife's certificate....?"* "I haven't GOT a midwife's certificate, there was no midwife at the birth" explained Alan. "Well then, there's NO BABY!" concluded the clerk.

"What do you want me to do, carry the baby down here to show you?" yelled Alan. But in the end, the clerk gave in and recorded Nico's birth, but only as a big favour, because it is illegal to go round giving birth like that!

Monika had a visit on the morning of the day Nico was born. They live in an isolated farmhouse in the middle of the woods, some distance from the nearest village. Monika was already in labour and she was washing some clothes when two old women from the village knocked on her door; they said they were just passing and came to say hello. By a strange coincidence, it was the birthday of the older woman, Maria, and she had been born on this very farm where Monika and Alan now lived. When the women realised that Monika was actually in labour, they were alarmed and said "But aren't you going to hospital?" "No, I'm going to have the baby right here where I feel at home! " replied Monika. "You are quite right!" said Maria. "We were all born right here at home too!" and then Maria told a lovely story about childbirth from the old days.

This was Maria's story: "One day when I was a little girl, a woman came by here on a horse with her husband: they were going home from the market at Gualdo. The pains had started as she was riding home and so she stopped here at our house. My mama was in the kitchen and as soon as she saw that young woman, she understood how things were. My mother had seen plenty of babies born, she knew how to help, too. "You go and lie down in the bedroom "she said, "There's Ninetta in there with her baby" . My big sister had had a baby just a few days before. So the woman got into bed with Ninetta and in no time at all, she'd had her baby. She called it after my mother, it was a way of saying thank you!" What a great story for Monika while her labour pains were growing!

I wish we had women again who could so confidently and casually invite us into their homes to give birth! And I wonder what that hospitable peasant woman would have thought if she could have seen the ten-metre flagpole now outside Ivo and Masako's farmhouse with the three enormous, golden fish-flags that flutter and glitter in the Umbrian wind to announce a Japanese birth in Vallingegno!

June
Franciscan paths

Sunday morning the sun was shining. Martin had left early with the horse Papu to go riding with his friend Massimo. He was actually torn over the decision: on the one hand, he wanted to see Massimo, who's a real horse man and was the one who taught him to shoe the horses, but on the other hand, it bugged Martin that this yearly ride to Assisi had been co-opted by the Region of Umbria as part of the official inauguration of the "St. Francis Way". This path runs through the countryside between Gubbio and Assisi, following the route Francesco took when he fled from his angry father to friends in Gubbio. People occasionally come past, quietly following the saint's footsteps on his adventure. The path passes our front door but of course it is only a little track through the woods. However, now the Jubilee Year is here, the authorities set aside funds, millions of lire, to "restore" the path. We held out against this because we were afraid that the spirit of St. Francis was going to disappear under the bulldozing, chain-sawing operation which has more to do with tourism and money than pilgrimage. They didn't like us and deviated the official path off our land, but the project went ahead, with its ugly signposts along the country ways, as if they wish they could turn the whole countryside into a medieval Disneyland.

Martin had decided to go: this ride was something they did every year and anyway, he thought he could stop at the river and swim with the horse instead of going all the way to Assisi and having to listen to the politicians' speeches. So he had set off.

I stayed home with two guests: an Australian woman and a Sicilian friend who teaches at Rome university but whose heart is really in his vegetable garden and whose dream is to go back and farm his great-grandfather's land down south.

He was here for the weekend to see how we live and in the morning he asked me "Can we walk all round the borders of the farm?" Of course, we are constantly moving all over this land, but actually the "beating of the bounds" was something I hadn't done since 1975 when the old owner of the farm took me all the way round before I bought Pratale.

I say "old", but Francesco Moricone was only 61 then. But I was 27 and in my youthful, arrogant way I saw him as an old man. His name and surname are the very same as St. Francis' real name!

The hillsides here are steep and on that occasion, I was almost sorry I'd asked Francesco to show me exactly where the borders ran because, despite his years, Francesco preceded me up the hills like a mountain goat: he had spent a lifetime moving up and down these fields. I wasn't used to country life yet and I lagged behind.

Twenty-five years have gone by since then and here I am retracing his footsteps up the hill, with more breath than I had then, Filippo and Annette following me. But now that I am the "owner" of this land, I understand the smiling silence of Francesco on that winter day. He was passing on the land to me, but the land contained his stories, his past, his whole life. These hills were so layered with meaning that he couldn't talk to me about them, could find no words to describe them, although he was a talkative man in the ordinary way. That cold, sunlit day we had looked at the pastures, the woods, the stream, the boundary stones in silence.

Filippo, Annette and I were above our woods: from there we could see the castle, the stream at the bottom of the ravine, and above the cliff, our neighbour's vineyard and his fields, which from there seemed almost perpendicular. The woods were thick and green and the yellow broom was everywhere in flower. But the feelings I had were quite different from that first time when I walked there with Francesco: now this landscape was teeming with life, stories pulling at my heart. I could see moments of past love and delight, days of summer work, times of earthquake terrors, people and animals I had loved and who had disappeared from these hillsides, the different face of the valley at other seasons of the year. How could I tell them all these things? I wanted to talk to Filippo, especially, because I knew I was living the dream he was still holding in his heart and that he wanted to hear how it had all come about for me.

I was surrounded by past images of myself, from ten years, twenty years ago, from last month; I watched myself as I moved across these steep fields on a hundred errands, I could see Martin up here in a tropical rainstorm at the autumn equinox two years ago with his shepherd's umbrella in shreds and the little dog trembling, my sister and myself as we picked rosehips with Clara sheep following us and the children playing and tumbling around with Clara. I could see a younger Antonio coming out of the wood with a sack of mushrooms at dawn...he was always our mushroom teacher. I could hear the farmer across the stream, who used to plough with oxen, as he screamed abuse at their placid backs to keep them walking straight. From up here, we used to marvel at the driving force in his words. He always used to be the first to start turning the earth over and, although everyone else used tractors, he was always the first to finish the work because on day two everyone else would be up at Rino's getting their broken tractor parts mended, standing in the blacksmith's queue impatiently, knowing that Chioni's oxen were moving slowly forward while they stood idle.

I could see the Christian knight Verecondo fifteen centuries ago standing just above the stream, preaching a new religion to the people celebrating the feast of Mars. He spoke quietly and passionately while the country people warned him to get off the land sacred to the Forest God Silvanus and I could see the stones flying. Here was where they lapidated Verecondo and buried him above the banks of the stream until five centuries later, the local people, carrying in their minds all that time the whereabouts of his bones, carried his remains up to their Christian abbey. Two hundred years later, a ragged man called Francesco of Assisi came past here, his hair wild and full of snow because the brigands on the other side of this hill had thrown him in a ditch where he lay back laughing and singing the praises of creation.

We came down towards the stream, walked back up the path, through the fields where the bundles of juniper kindling lay ready for next winter's fires, past the oak and ash logs that Martin cut and stacked for the horse to carry home. I am full of stories, the landscape is deeply overlayed with them, too many to tell unless there were plenty of time: we need long, firelit nights to rekindle all the lives that make up this landscape. And all that morning, as I was following the bounds, I could see, in my mind's eye, the spare little figure of Francesco leading me up and across his land, with his stories locked in his heart.

We came home and towards evening Martin came back on his tireless horse carrying the news he had heard at the country fair at Vallingegno: Francesco Moriconi died this morning. He was 86 year old. All those many years he'd walked these country places, his footsteps crossing and recrossing the Franciscan path.

Summer Solstice
The beauty of the world is a way in.

Walking along the trail I see bear tracks.
I hear them, turn around
to see the bear behind me -
It is a sunny day but still
there is snow on the trail.
I think how pleasant it would be
to die here and become bear fat,
to climb back up the avalanche chute
to look back across the muskeg...

All alone in sunlight, in snow,
in fur.

Gary Lawless, Sitka, Alaska 1995

A solitary path of inner searching has taken the place of religion for many people in our day. For some, the path starts from distant religious practices, from ancient philosophical systems or rituals borrowed from other cultures to act as guides in the desert, footsteps to follow in the absence of other comfort.

Originally, all religions were religions of *place* and they offered rituals for entering into the heart of the physical world, through the door of that specific place which had given birth to you or which had welcomed you. Nowadays, we Europeans have no living tradition of place and yet the truth is that the world puts all of us, at every moment of our lives, in exactly the right state to meet with the great mystery.

And this is the state of need, hunger. We are not self sufficient, we need food, water, warmth in order to stay here and so we have to go out and look for these things, gather them in. This need to gather pushes us towards the other beings and forces us into a particular state of mind, a state of awareness which alters our consciousness and opens us up, makes us vulnerable. There are many stories which tell what happens to the human being who goes out to gather berries, mushrooms, chicory, flowers, wild herbs, wood for the fire. These stories exist all over the world and I would like to tell some of them because they allow us to see what accumulated human wisdom has to say about the destiny of the person who moves out of her/ himself into the world to fetch what s/he needs.

In our European tradition, there is a wonderful picture of the state of mind of the gatherer in the myth of Persephone. The beautiful girl is drawn into a state of ecstatic contemplation by the beauty of a narcissus
she finds in the meadow. She forgets herself and that is exactly what happens to all of us if we go out picking flowers, blackberries, mushrooms: we wander from plant to plant until the landscape becomes just a movement from one hidden treasure to the next. We get lost, the world turns into a trail of shining berries.

And when we have completely forgotten ourselves so that we are only conscious of the other beings we are searching for, suddenly the ground yawns under our feet, a black mouth swallows us up. As Persephone's hand reaches for the narcissus, the abyss opens up and Hades, with his black horses, appears, seizes her and drags her down into the Underworld.

Pindar, speaking of the Eleusian Mysteries, says "Blessed is he who, having contemplated them, descends the path to the Underworld: he knows the end of life and its beginning".

The great religions, that is to say the religions of the "great" civilisations whose existence relies on controlling the masses, have turned their faces away from the physical world, deeming it an illusion or evil in nature, instead of seeing it as an invitation held out to every person, at all times, to come into the world's deep bed of being, that place where all the beings of the earth dream together. "One light shines from the multitude of the species" said Giordano Bruno. The proud beauty of the wild world, in all its manifestations, big and small, is not empty. The Underworld is the place of treasure, it is the source of the life of the spirit as well as of the life of physical forms.

In another European story, Little Red Riding Hood, the girl stops in the forest to pick flowers (although her name might suggest the gathering of that magic mushroom Amanita muscaria). Her fascination with the flowers makes her forget her mother's warning and she loses her sense of time, opens herself up to the beauty around her, and at this moment of vulnerability, that personification of the wild world, the Wolf, appears to eat her up.

This swallowing of the human being, this mixing of the wild animal and the human, is the result of the gatherer's movement out of her/himself and towards the other beings. The place itself devours the human being, makes her part of the community of all beings, because at last s/he has lost the armour we create around ourselves to keep our selves separate. Suddenly the hidden, inner dimension of the physical world is illuminated for us.

A theme of many Native American stories is the woman who goes out picking blackberries and meets a bear-man. She follows him to his home and marries him and has two children, half bear and half man. This story must have crossed the Bering Straits many millennia ago, because

there is a parallel story in the Russian lands. The version I have is called "Masha and the Bear". Masha lives with her old grandparents
and she begs them to let her go and pick berries with her friends in the forest. The grandparents tell her to stay close to the others or she will get lost. But Masha gradually forgets her friends as she moves from tree to tree and from bush to bush, and ends up in that dimension outside time, lost in the deepest part of the forest. Dusk comes and Masha sees a little hut so she slips inside it. But the hut belongs to a huge Bear and when he sees Masha, he is very pleased and says "You'll stay in my house/as quiet as a mouse/ and be my servant/faithful and true" And so Masha stays with the Bear for a long time, until she finds a way to trick him and get back home to the human community.

Entering the yawning abyss, being eaten, entering the hut, being married - they are all ways of being consumed, transformed, *completed* by the wild world. Losing your self and entering into intimate contact with the great self or soul of the world is the beginning of all ethics.

Here is the same story, this time from Arnhem Land in Australia, told by Wandjuk Marika of Yirkala. This story of being taken into the heart of one's place may date back forty thousand years. "In the Dreamtime, the two Wawilak sisters were travelling inland. One was pregnant and the other had a baby and as they travelled, they hunted and gathered and gave a name to each animal or plant that they ate. When they got near to the great Mirarrmina Waterhole, the younger sister started to feel the birth pains, so they stopped to eat and rest. They didn't know that the giant Rainbow snake lived in the depths of that pool. But when they tried to cook their food, each plant and animal jumped out of the fire and ran to the waterhole and jumped in it: they feared the Serpent. Thus with the crab, the yams, goannas, frill-necked lizards, rock pythons, seagulls, sea eagles and crocodiles - they all disappeared in the waters of the well. The elder sister then went into the waterhole to search for some paperbark to make a cradle for the newborn baby, but some of her menstrual blood fell into the water. The great Serpent under the water smelt that blood "Where does this blood come from?" he asked and he crawled out of the well spitting water into the sky, where a black cloud formed. It was getting dark and the two sisters said "Where does this cloud come from?" but they were unaware of the Serpent. He was furious, and rose onto his tail and was so huge that he reached from the earth to the sky. A terrible storm broke and as the rain poured down, the sisters finally saw the Great Serpent. The elder sister danced and shouted to frighten him away, but the Serpent ate the younger sister and her baby. Later, he regurgitated them and they came to life again."

It is interesting to see how all the animals and plants were aware of the Serpent's existence and ran to plunge into his waters; only the human beings were unconscious of the spirit of the place, unaware that they also belonged to the place and were part of it. They were unable to recognise their own bodily fluids as messages to the outside or understand the source of the rain which allows all life to flourish. But they had taken the first step in hunting and gathering the other beings and thus had signalled their state of need and dependence. The Serpent contains and transforms them and this is an image of the wild world breaking into human consciousness; all those who inhabit the Australian places, aboriginal or not, have to come to terms with the physical place and the spirit of that place, the real Self.

Deep inside us lives the actual place we inhabit and are part of: the earth dream is one, but the way in is along familiar paths. If I go out into the night here, I meet wild boar, badger, fox, porcupine, deer. In the daytime, I go amongst jay, buzzard, crow, heron, blackbird, whip snake, viper, lizard, hare. The shadows of what remains of the Umbrian Forest surround me and I recognise other parts of my self. The lichens, for example, seem like a small, marginal part of the landscape until I go out to collect them for dyeing. Then I begin to see their enormous variety of shape and colour, how they grow across the trees and stones and dead wood, how they crowd around the banks of the stream, which trees they prefer, how they grow and fall, move and reproduce. Suddenly I see the lichen as the centre of the world, perceive the me-ness of it. That is what the Australian aboriginal culture celebrated in their increase ceremonies, when they imitated the appearance, the movements, the life of the green caterpillar, the kangaroo, the scorpion, the emu, the honey-ant, the spider, the wild plants. They danced and sang to feel the selfhood of every being. The Aranda song for the yelka plant (Cyperus rotondus), given to the Australian scholar T.G.H. Strehlow as a personal gift, describes the plant as the singers see it in a good season, after copious rain, with its flowerhead red and heavy and the roots like many fingers, the plant at its most perfect. If you want to find food, another being, you have to *become* that being, let go of your human armour, open up to the quintessence of that being, live the life of the other.

Gather and be gathered: we have a great need to get back to the most ancient religion in the world, walk along the paths of our places, lose our sense of time and lose our selves in the other beings. If we think in terms of our relationship with place, we have to ask what can foreign religions like Christianity or Islam, imported from the Middle East, or Buddhism, imported from the Far East, have to offer us people (of *whatever* origin) who now inhabit what remains of the great European forests? Worse still, what can a supermarket offer us?

Can we give up some of our comforts and pay for the renewed vision with the old, hard, precarious way of life? Not many people will do that, but it is important to be aware that every step that distances us from a life of direct, felt, dependency on the other beings, takes us further away from the life of the spirit. So it is better to depend on your legs than on petrol, better to depend on fire than nuclear power, better to eat the wild plants than the cultivated ones, better to farm in the old ways than in the industrial mode, better to make than buy.

Nature does not reveal the intelligence behind her beauty to those who merely admire her, but only to those who are prepared to go towards her, in their need, taking little and losing themselves along the way.

July
Miranda and the wasps

My niece Miranda has arrived. Maybe it's because of some affinity deriving from her Italian name: the fact is that she comes here every summer, despite her horror of insects. Now she's fifteen and so she doesn't let out the heart-stopping screams that she used to give vent to as a child at the sight of a wasp, but I quite often come across her staring at something small with an expression of deep disgust. It means she's found another scorpion in the bathroom or what she calls "a double wasp", by which she means an ichneumon, who's trying to fly into her bedroom to build its nest, or maybe a praying mantis, looking up at her with its questioning, intelligent stare. "*Why* are there so many insects in Italy?" she wails "And all of them *outsize*?" Ah well, this year I have done some research and so I have some bad news for poor Miranda: if the world population of insects were divided by the total human population, there would be somewhere in the region of *two hundred million* insects each! (The calculations were done by the entomologist C.B.Williams) So there's no use getting upset about a single scorpion when there are another 199.999.999 insects lying in wait just for you!

Here is Miranda helping us dance on a big heap of lentils piled in the courtyard. It's a primitive way to thresh them but it's fun and there are plenty of us. In the corner of the courtyard is our "poor man's swimming pool", a stone-walled pool of fresh water just big enough for three of four people to lie in when the day gets too hot. A solar pump pours water from the spring in the woods into this pool every morning and in the evening we water the vegetable garden with it. During the day, the local wasps use it to drink from and to carry water off for their larvae. We put a saucer of water on the wall of the pool so they will drink from there instead of landing in the pool because at midday there is an amazing number of wasps flying backwards and forwards. It is not easy to keep one's calm in the midst of this cloud of wasps. Now, while we thresh the lentils, I watch Miranda. As she dances, she also keeps a wary eye on the wasps and weaves and ducks constantly to avoid them. I try to imagine what it must be like to be a wasp and have to negotiate a dancing girl in a miniskirt who seems *intent* on collision, just asking to be stung! It probably takes thousands of flight hours to learn how to deal with this sort of obstacle! Sure enough, she does get stung in the end and her cries are louder than the African singing we are dancing to.

So I doubt whether persuasion or example will ever convince Miranda to live peacefully with the insects of Pratale and I am always honoured that she feels being here is worth all the terror!

It *is* difficult to contemplate insects we are afraid of without feeling threatened. Yesterday I was lying in the water feeling the delicious cool after working in the fields, but one wasp after another kept landing on the surface of the water, close to my naked skin. Then I began to notice what beautiful shapes, like flowers, their little shadows made on the water and how it felt to be fanned by their wings as they flew close to me. Then my mind shifted and the whirling yellow bodies suddenly seemed like a miracle: no longer concentrated on defending myself, I saw their vitality, their amazing capacity to float on the water and drink and the magical parabolas of their flight for what it is - the exquisite perfection and grace of nature. Insects are very, very ancient: they were here 230 million years ago, before the mammals existed, and wasps are the *ancestors* of ants, which seems extraordinary! They have had a lot of time to hone their ways to perfection and are perhaps waiting for us to do the same and stop thrashing about and swiping at them!

Two years ago, during the summer, hornets appeared in the elderflower tree near the old house. There was a big hole at the base of the trunk, a perfect place for their nest, and at the beginning there were only one or two hovering about at any one time. They were too close to the house and we knew that if we left them in peace, their numbers would soon explode. So Martin declared war on them. The battle raged for days. The insects were amazingly resistant and in the end, Martin had to wall up their hole with cement. The next day, there was a survivor back, already digging in the cement trying to open the hole again. I knew we had to stop them, but when I thought about the old elder tree, always the first tree to come into leaf in the spring, those ferocious yellow hornets with their determination and strength felt as if they were a true part of the tree, an added richness in its life, just like the white flowers, the black, juicy berries and those pale green leaves with their strange, narcotic perfume. Tradition says that the elder is inhabited by Mother Elder, the divinity of the tree and in many places in Europe it is a magic tree, which you must never cut without asking the Mother's permission. And supposing she herself had invited those hornets?

Yesterday, Miranda found a tiny scorpion *under her bed* and she was really horrified because she couldn't help wondering where the mother scorpion was. It was no use repeating that this kind of scorpion isn't dangerous, that they sting with great reluctance and it is only like a bee-sting even then. It did no good to try and soften her feelings towards scorpions by telling her how the mother carries her babies on her back, ranged in two lines: Miranda just wanted to hunt her down and kill her. So we had to tell her the local story about scorpions. A few years ago, someone came down from the city and was staying at a neighbour's place in the valley. In the night he woke up and found a scorpion near his bed. He didn't know what to do, so he imprisoned the insect in a glass jar and left it on his bedside table. In the early hours, he woke up again and looked to see what time it was by the clock near the bed. Then he noticed that on the wall near his pillow was a whole semicircle of large scorpions, all looking in his direction as if to say "What have you done with our friend?" The visitor threw the prisoner scorpion to freedom out of the window and then picked up his eiderdown and went down and slept on the terrace. Next morning, he told his hosts what had happened and in Valdichiascio the idea took root that it is better not to bother scorpions. It is true that they are elusive insects: if you poke them with a twig, they try to get away rather than sting...all they want to do is get back into the shadows and mind their own business.

Ah lovely Miranda, who comes in the middle of the great Italian heat when the earth is pulsing with insect life! Will she always see them as the enemy or will she one day suddenly see their clicking, whirring, glittering presence as part of life's rich pattern?

August
Titania at the gate

In August, Pratale was full of people and just making enough bread every day was an undertaking. Random groups formed as people who didn't know each other came and went and sometimes it seemed as if there must be some hidden reason why these particular people should have met.

Sometimes I think "How on earth will these two ever get on?" Yet I've had to acknowledge there are bizarre, illogical ties linking the people who turn up. This year, my brother and sister came. It's rare that all three of us are together nowadays. The next visitor was a Hawaiian woman and as she walked in, I was stunned to see someone who looked just like our dead mother: it was as if Jane, Tim and I had evoked her by our coming together in one place. And when was the birthday of this warm, soft, clever woman ? The same day as our mother's.

The day after Nimmy arrived, we all spent the very early morning blackberrying because we wanted to make apple and blackberry jam. So after breakfast - sheep's milk yogurt and wild black cherries - we sat round under the mulberry tree to peel the tub of apples. When we do jobs like this, someone usually reads aloud - we've got through a lot of good books like this. There happened to be an actor with a lovely voice here and I went and pulled a play at random out of the library, "Do you feel like reading to us while we peel?" I asked him. Everyone looked up expectantly. "Do you know the play?" I asked. In reply, Steve laughed and opened the book at page one where it says who first performed the play and in which theatre. I read his name at the top of the list of actors. "That play was the most fun I ever had on stage!" he said.

When a group first meets, there's a sort of honeymoon: we are all very *nice* to each other. Then the rougher sides start emerging and the waves begin. Two of the big kids get together and the third is angry and jealous, couples start arguing and try to find allies amongst the group, the little children scream with joy or fury during the hot hours and wake up those who are trying to siesta, the people who work in the fields at dawn come home for breakfast to find the yogurt's all been eaten already and they feel resentful and unappreciated. And then, over one or other of these incidents, the group recognizes itself, crystallizes and discovers it has a heart and can act in unison. Everyone manages to laugh because someone says *"I'm* the Guardian of the Yogurt and from now on *nobody's* touching the yogurt, right, 'til all the workers are here!" Suddenly, instead of twenty random people, there's a group working and eating together. It is interesting how rage and irritation are often better catalysts than smiles and talk.

Martin repaired the courtyard gate for the tenth time: he put a new latch on it which we hoped was donkey-proof. But the donkey Titania, she of the huge, pregnant belly, managed to figure it out straight away: she knocked the latch up with her nose, slipped her rubber lip through the gap and slammed the wooden gate open with her head. The baby, Jip, who was only one, screamed in horror as Titania's vast head suddenly appeared at the gate. But she didn't want to come in and eat the geraniums and it wasn't designed to demonstrate her skill with locks, she just wanted to stand there, enjoying the company and observing the humans.

Maybe she perceives this interesting, flowing human work, so energy-consuming and swirling with undercurrents, which is the building of a common heart. Michael painted a watercolour of the courtyard from Titania's vantage point.

What was her view of things, as she stood there quietly watching us, hour after hour?

She could see the stone bath we built in the corner of the courtyard. It's a Japanese bath, a poor man's swimming pool, filled during the heat of the day by the solar pump, for watering the vegetable gardens in the evening. Sometimes there were three or four little kids splashing each other or a group of large teenagers making the heaving water level rise alarmingly while they soaked and discussed exams, beaches, plans for life, or the adults talking who's making supper or how to sort the world. Did Titania like our truce with the wasps who came to drink from the bath? We left them a shallow bowl of water on the wall and they needed no explanations but drank there, minding their own business and leaving us in peace. Even the small kids took no notice of the stream of wasps landing, drinking and flying off again.

In the middle of the courtyard, under the shady branches of the mulberry , Titania could see the table, getting longer each day as people arrived and we added more planks. The tree was full of climbing children and at the table there were always people writing, playing chequers, chopping vegetables for the next meal, and talking, talking, talking, sometimes in several languages and with their hands too. What is this human talk thing? Like a self-replicating virus that we have caught.

Under the vine, Titania could see a woman playing a violin and another woman playing the Irish whistle with a boy on the digeridoo. And on the other side of the far gate, she could see two more donkey ears - Otello, come to listen to the human music.

When the day of the big eclipse arrived, the group was up early, busy building an extraordinary contraption under

Titania's quiet gaze. It was an ambitious piece of equipment constructed with three ladders, two tall ash poles, several blankets, plenty of string, nails, binoculars, a sheet and a lot of advice. We were trying to follow the instructions of a scientific friend who'd already left for Cornwall and the total eclipse. The contraption was designed to project the image of the eclipsed sun onto the sheet hanging off the roof of the house. But we couldn't fix the nuts and bolts so as to give a clear image and in the end, my brother Tim sacrificed himself for the community by standing in one spot on the precarious ladder, holding the binoculars steady as the sun began to disappear.

As that strange hour advanced, we saw that all our efforts were unnecessary - like the dark glasses being sold all over Europe - because Nature provided thousands of tiny images of the diminishing sun under every leafy tree. It was magic to see the circles of sunlight changing slowly into crescent-moon shapes. None of us had ever realised until then that the splashes of sunlight under trees are not random patches of light filtering through the branches, but actual pinhole-camera images of the whole sun. Under the blue and cloudless sky, we felt the cool of the moon-shadow and we wandered between the mulberry and the silent fields, where the birds and the sheep and donkeys seemed to have fallen into a sort of torpor.

The Sun has to do with the hard, rational, established structures of authority and power and the Moon has to do with the flow of feeling. The last total eclipse ('59 in England, '61 in Italy) preceded the Sixties, a period which saw the dismantling of worn-out hierarchies. Now we were standing on the threshold of another period during which the feelings of the heart will dissolve outdated power structures.

During that eclipse, some of us - especially the children - felt sick. In England, a good many pigeon-fanciers lost their homing pigeons and people who were at the beach during this coming together of the Sun and Moon felt the sea heaving. People who were in Turkey felt the earth shake in a tragic "dissolving of structures" as houses collapsed in the earthquake a few days later, and the shaking spun round the world for several weeks...in the Taiwan quake, the epicentre was at a place called Sun and Moon Lake and the quake in Greece centred on a village called Metamorfosis. These were all physical manifestations of the forthcoming inner changes announced by a major eclipse and although we didn't hear anything about this from our newspapers (we live in reductive times), we all felt how significant that strange hour was for sublunary life as we moved quietly round the courtyard and stared at the little moon-shapes reflected on our bare feet.

All this Titania could see from her place at the gate.

September
The lost lamb

The lamb called Ultima disappeared two months ago, in June. We were all getting supper when we heard Martin call "Hey! There's a lamb missing!" It was dusk when he went to shut the sheep up and it was late to go looking for lost animals. But we couldn't go to bed with that on our minds, so we left the supper preparations and everyone went in a different direction. We called each other "Have you looked above the woods?" "No, I'm on the rocky place" "Has someone been up to the Top Field?" "Melissa's up there" "I've been all over the olive grove" "Have you looked in the fields under Luigi's house?" It's a good thing that these sheep are white, at least they show up in the dark. There isn't an inch of the land that we don't know but it's also true that the fields are dotted with trees and rocks and there are all sorts of cliffs and dangerous places where we have lost animals in the past. We searched all the fields as the night grew darker but in the end we had to give up. Clara's lamb, the last lamb born this year, that we were going to raise, was lost.

The next morning, as soon as we had fed the animals, we started searching again, but even in daylight, we could find no trace of Ultima. She seemed to have disappeared into thin air. Clara bleated all over the pastures, but there was no reply.

Now it's September, and we had resigned ourselves to the loss, but late one evening last week, Martin was working in the garden when he heard bleating which seemed to come from the woods on the opposite side of the valley. Without alerting anyone, he ran down the hill, crossed the stream and made his way fast through the undergrowth and up the other side. Yes! He'd heard aright! He could see something white moving through the trees. Running as fast as he could through the wood, he managed to throw himself onto the lamb and catch hold of it by the fleece: it was Ultima indeed! He heaved the lamb onto his shoulders, in Good Shepherd style, and then he looked around to find the way back down to the stream. Ultima was quite heavy - life in the wild obviously suited her. But now, whichever way Martin looked, he seemed to be hemmed in by brambles as high as a man. All he had on was a thin Tshirt: in the fury of the hunt he hadn't felt the thorns, but now in the darkness it seemed impossible to find a way back out of the woods.

I was in the house when an Israeli girl, who was staying here at the time, came to doorway and said "I can hear someone in the woods, shouting for help!" We both listened. It was true! We started down the dark hill and reached the banks of the stream. Sure enough, it was Martin's voice from somewhere above our heads. "I've found Ultima! But I can't get back down! You'll have to get sickles and come and cut me out of here!" "Right!" we shouted back.

Back at the house, I told Neil what had happened. It was really black night now and our tools were pretty poor: we found two rather blunt sickles and a torch which seemed about to fade out entirely, plus two asthma sprays because both Neil and I are spray-fetishists. We picked our way back down the hill, tripping and falling into holes as we went. We crossed the dark stream, soaking all four shoes. On the other side, we found we couldn't actually walk on the ground at all because the undergrowth was so thick that we were really walking on matted plants about a metre above the ground. We could hear Martin calling from a place which seemed to be up a precipice over our heads. We started laughing because the whole situation seemed so hopeless. We turned the precious torch on and in its faltering light we could see a few square centimetres of undergrowth. We tried slashing at the undergrowth, feeling like Che Guevara in the jungle without a machete. Come to think of it, Che suffered from asthma too and how I felt for him now! Neil, usually the epitome of your fiery optimist, stood there in his wet shoes holding his blunt sickle and suddenly felt despair. "Martin, you'll just have to sleep there. We'll come back tomorrow when we can see!" he shouted.

"You must be joking!" replied Martin, "I'm not that far away, what's the problem? You've got sickles and I'm stuck here with this sheep on my shoulders!"

I agreed with Neil, but I kept slashing at the undergrowth so as not to depress Martin: I hoped the sound would keep his spirits up, but my actual progress was almost nil. I tried both sickles: they were about equally blunt. I managed to make a clear space of about a metre and a half. Neil was still acting paralysed.

Suddenly, Neil came out of his trance, picked up the sickle and started slashing energetically at the undergrowth. In a few minutes, he'd cleared several metres of path. Martin shouted encouragement from his precipice and Ultima bleated. We worked for hours, but by now we had got into a rhythm. We were aware of time passing. By now the others would have gone ahead and eaten supper without us. When we reached Martin and Ultima, it must have been about one in the morning, but the Good Shepherd had lost all sense of time. "Is it supper-time yet?" he asked us.

Later that night, with Ultima restored to her flock and the three of us sitting in front of plates of hot spaghetti, I said to Neil, "So what happened to you down there in the woods, all of a sudden, when you went in there like a freedom fighter?" "Well, to tell the truth, " said Neil, who is an anthropologist, "down there in the dark, in the middle of all that undergrowth, I suddenly remembered the terrible wounds from wild boars which the Marsh Arabs showed the author of the book I'm just reading. They rip your belly open, you know, with those tusks. So I suddenly thought, here I am and I can't see the wild boars, but they can see me. So it seemed better to get on with it instead of standing there with that thought in my head!"

"I see, Neil! You mean you left me behind you as alternative bait!" I said, but anyway the adventure was over now and we could laugh at our fears. The main thing was that Ultima was back with Clara in the stable.

October
Three Sardinian stories

"One night there was a big storm. I was little, but I remember it. We were all scared, at our house. There was a terrible high wind, it was carrying things away in the dark, thunder and lightning, rain beating on the roof. It went on all night. Nobody could sleep. In the morning, there were lots of huge branches on the ground, damage to the houses in the village, the sheep enclosures broken. My father went out early with my uncle to see what had happened."

This is one of the stories Luigi told me about his childhood in Sardinia. We were sitting on the hillside, overlooking the field where his large flock of sheep were grazing. The fiery red autumnal woods opposite made a background to the stories I heard. Sometimes, nowadays, sitting on the pasture in breaks from field work, I remember those scenes Luigi painted, stories from another world, not so removed in time but very distant in spirit.

"There was terrible damage, but the most dreadful thing was an oak tree at the end of the village. It was in front of the witch's house." "The *witch*?" I asked.

"Yes, she was an old woman who gave out medicines, she made them. I was afraid of her, I never went near that place. But my mamma went, all the people went. That morning, I remember, I went with my father to see that enormous big oak tree in front of her house that the wind had uprooted in the night. That was the only tree that had been uprooted, and it was *standing upside down*! With its roots in the air! Everybody in the village stood round it, in a big circle, nobody spoke, there was silence."

I imagined that strange scene and remembered that the tree with its roots towards heaven and the foliage on the earth is a very ancient magic symbol.

"Then a man went and stood in the doorway and peered into the house and he started shouting: the witch was dead! She had died in her sleep, quietly, in her own bed! She was very old. I was little, but I remember that morning!"

What an honour for that elderly country witch, to be accompanied in her last journey by such a majestic oak tree! And how strange to have lived through a childhood in that archaic world. The stories that we told each other, Luigi and I, in those years, were like two foreign currencies, so far apart were the worlds we had lived in.

Another morning, he told me a story about the journey for the *transumanza.* "We had to take the sheep up into the higher hills, for the summer. It was a long way: my brother and I took over a week to get there. We were passing through a little mountain village - it was just a single street, a dirt track with a few houses and a bar. We went through with the flocks, I was behind and my brother was in front. Then I saw that one of the rams - we had five or six - had caught sight of something and stopped in his tracks. We were level with the village bar and the door was made of glass. That ram had seen his own reflection in the window and he thought it was a ram from another flock. In a flash, he charged that ram and broke the door: the glass crashed into smithereens! The men who were drinking in the bar suddenly found that ram in their midst. He had stopped now, nonplussed, he couldn't understand where the other ram had gone. So he turned and instead of going back out of the door he'd broken, he ran through into the other part of the bar, where the little shop was with another glass door that gave onto the road. From the shop, the ram saw his flock again and so - crash! - he burst through that door too and ran back into the flock. Everything happened so quickly!"

"And what did you do?"

" We laughed until we cried - me, my brother, all the men in the bar, everyone in the whole village! How we *laughed* that time! But we had to pay for both the glass doors."

The great thing about this story is the thought that all the people in that remote village would surely know sheep well and so the swift reasoning of the ram would have been clear to everyone! It could have happened to anybody, but fortunately it had happened to a stranger and so it was he who had to pay for the joke!

But the story which sticks in my mind like a vivid dream is the one about the wild horses for the threshing. The scenes in that story seemed like a vision from an old, lost world. Luigi told it one day when a young Belgian boy had turned up at our house with the Flemish bagpipes. He began to play outside the farmhouse, and after a few minutes, the donkey Otello came up the path from the bottom fields. He came closer and closer to the place where the music was coming from until he was standing right behind the boy, who was suddenly startled to see the tips of two long ears, one either side of him, as if Otello wanted to get the full stereo effect at close quarters.

I was as astonished as the bagpipe player, but Luigi said "All horses love music". And to illustrate this truth, he told us how they used to do the threshing in his village when he was small.

"Everyone got up before dawn. Five or six men would go out. One had the accordion, one was the singer and the other men had ropes. It would still be dark when they got right up into the mountains. Then they'd sit on some rocks and wait. The one with the accordion started playing and the man who knew the songs for wild horses would start singing. Sometimes they had to wait for an hour or more, but in the end the horses always came, gradually appearing out of the mist and getting closer and closer to listen to the music. When they hear music, horses lose their fear - you can do what you like with them. When the men with the ropes could see six or seven horses, they'd look at each other and then get up very slowly. The men who made the music never paused and the men with the ropes would rope up the horses and then, very quietly and slowly, they'd all come back down towards the field above our farmhouse, singing and playing and bringing in those horses."

"On the field, the corn was piled around a tall pole, well set into the ground. They tied all the horses in a line to the pole, so as they would circle round it, led by a man on the outside. All day they went round and round the pole, while the men forked the corn under their hooves. The man playing the accordion could only have a rest when a neighbour took over, playing on his own instrument until he'd got the tune because the music must never stop. The singer who'd sung the horses in to be caught had other men who'd take over the singing and let him rest sometimes and drink. It was a very long day and hot! And then, in the evening, when it was almost dark and the threshing was all done, my father untied the horses, one by one. At that moment, everyone who was on the field would stand back out of the way - and then the men who were playing and singing would stop. It was as if those horses woke up and became wild again and they'd gallop away into the night. We were poor, we couldn't have kept so many horses and fed them all year round just for that one day of threshing."

They were fascinating stories. The temptation to turn Luigi into a romantic figure was very strong. He did what he could to avoid it. Many times, I heard him say "It'll rain tomorrow!" or "Today it's going to snow!" and when someone asked him "How can you tell, Luigi?", expecting him to come out with some old country weather lore, he would reply "They said so on the radio".

November
The olive harvest

"For St. Clement's day / Put your oxen away / On the twenty-fifth / Comes St. Catherine's / End of sowing time / There's olive trees to climb" . This is a translation of the proverb that the old people round here repeat so as to remember the right date to start picking the olives in November. Down on the plain near Perugia, they'd already been picking for several days and even here round Gubbio, the olives seemed further on than usual because of the warm October we'd had. But we were so used to waiting for the twenty-fifth that we hesitated to begin before that magic date; then the rain set in and, in the end, it was exactly 25th November when we assembled our buckets, ladders and nets and began.

The picking team this year consisted of me, Martin, our friend Teresa, her children Pietro and Anna and their friend little Viola. Sometimes our kids Ben and Camilla came to help and our Australian friend Colin came too some days. We made our ritual telephone call to our eldest daughter Melissa in London to commiserate with her on missing the olives: she picked so many basketfuls all through her childhood that she says she is delighted to enjoy entirely olive-less Novembers now. So really there were just three of us, because our kids and Teresa's usually had pressing business elsewhere. We were comforted to read a story in the local paper about olive-picking in the old days. "Granddad used to say, *If them olive trees had to wait for these 'ere youngsters to do the picking , they'd shrivel up and die of old-age, waiting"*. Ah, so it's *not* just our kids, it's kids through the ages!

We spread the net under the first tree, Teresa and Martin climbed up and stood on the branches and I climbed up the ladder. Teresa's huge puppy Bea was tied to a nearby tree because it seemed that she might go a bit too far when playing games with passing hens. Our two small dogs positioned themselves on the net, sitting upright on guard underneath the shower of falling olives. They didn't have to wait long. After about twenty minutes, we saw the three donkeys strolling casually in our direction. They stopped at a short distance from us, put their heads down and grazed closer and closer to the edge of the olive net, keeping a wary eye on the dogs. Vispo and Pisirù obviously had a private rule about the exact whereabouts of the invisible safety line separating the olives from the donkeys' mouths. As soon as the donkeys crossed that line, both dogs leapt up, barking "THIEF! THIEF!" The donkeys galloped off. We were grateful because it is so time-consuming to have to climb back down and tread carefully around the olives on the net to chase donkeys once we are up the trees.

But the noise had an undesirable side effect: it caught the attention of the sheep, who hadn't noticed yet that the humans were picking something. They came running to help. Sheep are far worse than donkeys, because donkeys eat slowly and make a lot of noise as they crunch the hard olive pits, thus giving themselves away so that, even without the dogs, we'd notice them. But a sheep is a fast-food animal and can eat in thirty seconds what a person has picked in five minutes. Worse still, there's always some courageous old hag of a ewe who knows perfectly well that if she doesn't run away, the dogs won't know what to do about it - they are only small dogs, after all, not wolves! Sure enough, one warrior-sheep stood her ground and stamped her foot threateningly while she chewed a mouthful of olives. Vispo looked up at us, helplessly. "What am I to do with this impertinent sheep?" We jumped out of the tree, yelling, and the sheep ran after the rest of the flock. Vispo and Pisirù retrieved their guard-dog dignity. Bea was a better deterrent because she is a large griffon dog and the sheep were nervous, unsure whether this foreign dog were tied up or not.

All our neighbours were picking at the same time, so the olive groves were full of voices and when we met, all the talk was of how many sacks picked and how much was still on the trees. Our elderly neighbours Rosa and Antonio with their relations picked two thousand kilos in five days: we felt we had a long way to go to reach such high speeds! But by the end of a week, we'd finished too and we spent a couple of evenings kneeling on the floor of the barn with the radio playing, picking all the leaves out of the piles of olives, moving carefully so as not to squash any.

Then came the day for taking the olives to the oil press. Martin set off in his tiny three-wheeler Piaggio truck loaded with six 50 kilo sacks of olives and the empty demijohns. Teresa and I watched him anxiously as he drove up the steep hill: the front wheel looked as if it would lift clear off the ground at any moment. Then we set off in Teresa's car, piled to the roof with more sacks plus her huge dog, to fetch her children from school and follow Martin. We had an appointment at a bar along the road to Perugia, because the only one who knew how to find the small country press was Martin.

But fate was against us. Halfway down the hill to Perugia, we found the whole road blocked by the local Fire Brigade who were trying to pull an enormous lorry out of a ditch. The children were delighted, but a traffic jam was growing in both directions, towards Gubbio and towards Perugia, composed of people trying to drive home for lunch. Everyone got on their mobiles and demanded sympathy from their mothers. We were worrying about our meeting with Martin and his one thirty appointment with the press. Once the olives are picked, it is vital to press them fast, so the acidity content doesn't rise and if you miss your turn at the press, who knows how long the olives might have to sit in their sacks? Anna and Pietro, squeezed between Bea and the sacks of olives, watched the firemen's crane and shouted out of the window to their teacher, Maestro Aldo, stuck in the jam just behind us. But Maestro Aldo had a jeep and after a while, he drove off across the fields, impatient for lunch.

When we were finally liberated, we drove fast to the bar and found it shut: there was no sign of Martin. He must have gone on to the press and decided we would just have to figure it out. We drove on into Perugia and asked around for the press. Nobody had ever heard of it. It was lunch time and the streets were pretty deserted. Finally a man with three missing fingers pointed the way with his thumb and we found ourselves in a bizarre and hilly suburb with a curious mixture of old stone farmhouses and brand new concrete villas interspersed with surviving countryside. Right at the top of the hill, Teresa spotted a beaten-up piece of cardboard nailed to a tree on which someone had written FRANTOIO in uncertain handwriting. It seemed like a secret place for connoisseurs only. We turned down a drive lined with cypresses and to our great relief, we could see an old farm outhouse with clear signs of the ceaseless work that means olive oil is being processed. Men were busy unloading sacks, all sorts of filter equipment was piled around the doorway and a strong smell of fresh oil and *sansa*, crushed olive residue, reached our nostrils. Martin's empty truck was parked in a corner.

Our olives were already in the mill and Martin was there watching the pressing procedure. (Old Francesco used to say "You have to *watch*. It's not I don't trust the owners of that press, but they sent all three kids to university, see?") In the room where the huge stone was turning, a circle of farmers stood round the tap from the centrifuge where the golden-green oil spurted into a churn. They were intent on comparing yields, good years and bad and the merits of different kinds of olives. It was warm and cosy in there, the wood stove was burning and a table stood next to the stove, with wine, bread, garlic and salt. The *padrone* took the children by the arm and made them slices of hot bruschetta. What tastes better than bruschetta with new olive oil?

December
Calogero's lake

"Hey, you know what? There's a lake down there!" Martin was climbing back up the steep slope below the pasture, walking through the dry, frosty grass. He'd gone down to the stream, Cerquetelle, which runs along our valley floor to see what the flow was like after the long winter drought. He'd just finished a piece of work that had taken him two months, clearing this old piece of pasture which had begun to look like a wood. We had realised that the trees and undergrowth so close to the old farmhouse gave the fox an easy run all the way from the deep woods almost to the hen-house door and that we'd do well to keep the area around the farmyard clear, as the old occupants of this farm used to do. Now he'd finished the work, the pasture looked like a park, with big oaks here and there and tidy stacks of firewood for next winter piled at intervals across the hillside. The view across the valley was opened up so that we could see the castle, which belongs to a prosperous southern doctor, and Calogero's farm down the track from the castle towers.

"A *lake?* What do you mean?" I replied. Below this slope, the stream is narrow and the water is never enough for swimming in the summer. Martin laughed: "I think it must be Calogero's famous lake! He's finally managed to make it!"

Sixteen years ago, one misty August morning, we had heard the rumble of a bulldozer on the opposite side of the hill. Afraid that someone was intending to build over there, an exploratory party set out, following the path up through the woods towards the place where the noise was coming from. I remember there was a boy from Catalonia, Esteban, and an Australian woman called Cassandra and a rabble of small children. When they got as far as the castle, Martin met

Francesco and asked him what was going on. Francesco laughed and said "It's Calogero! He wants to make a lake up there. A lake on a hilltop! It'll never work - how can water stay up there?"

He was right. The lake filled up with water from the spring, but the field below the lake gradually became saturated with seeping water and the earth started slipping down on top of the underlying clay, carrying stones and saplings with it towards the cleft in the hillside where water drained into the stream in winter. Every so often, Calogero would fetch the bulldozer back and dig the lake out deeper: it had become a sort of challenge.

The big shot at the castle took to photographing the erosion every time he came on holiday and in the end, his battle against Calogero became his raison d'etre. We found ourselves in an unfortunate position, agreeing with him about the damage but for different reasons: we were sorry to see the damage but not for considerations of private property. He came to ask us to be witnesses when he sued Calogero and we tried to convince him that you just can't sue your neighbours in the country - you have to go and reason with them. We tried to reason with Calogero ("Look, even that big oak is keeling over now! Maybe you should plant a row of trees down there to hold the earth?") but he maintained things were fine ("That oak's been there ever since my grandfather's time!"). The castle owner shut off the track leading to Calogero's bottom fields, so that he couldn't get down to them with the tractor. There was a lot of grinding of teeth.

Time is the only winner. Over the years, the castle owner wearied of his battle and we wearied of reasoning, but the water continued to follow its nature and seek its lowest level. Every time it rained, an incredible landslide of mud and stones crashed down the hillside from Calogero's field into the stream, the noise booming across the valley.

So this is how it came about that a dam of accumulated mud and stones rose across our stream, so that now there was a little lake just like Calogero dreamt of, only it was four hundred metres lower down the hill! I went down to have a look. The new lake was in a place we never go past, because the valley sides are steep and covered in trees and undergrowth at that point, and so when I'd struggled through the vegetation and looked down at the streambed, it seemed like a miracle! There, indeed, was the lake! It was covered with thick ice but maybe in summer we can bathe there.

Water tends always towards the abyss, as the I Ching says!

Winter Solstice
Who sings our lost rivers?

How does it come about that the countercultural move to reinhabit the land, which has been going on in the industrialised parts of the world since the sixties, is articulated by poets rather than by agrarian experts or sociologists? The migration from city to countryside was one of the cornerstones of bioregionalism, a label coined to draw attention to our place in the natural world, and at the same time to leach power from the artificial borders of man-made counties and states. In the case of the United States, for example, many of the state lines were drawn with a ruler, cutting across mountains and rivers as if these physical realities were unimportant. But it is the mountain range, the plain, the wetland and the river which says everything about the vegetation, the kind of wildlife and the forms of human culture that arise in that particular place.

Here is a good example of a bioregion: until the 1700s, there was a unique and geologically recent ecosystem in the east of England called the Fens. Twelve thousand years ago, the glacier which covered the land as far south as the river Thames began to retreat northwards. The level of the sea rose and the waters of the Wash flooded the Fens basin, which was made up of chalk hills in the east and a Jurassic calcareous escarpment to the west. Aquatic plants grew in the shallow waters and the dead leaves from these plants gradually reduced the depth of the waters until various kinds of reeds were able to grow there. Thus a peat bog formed above the impermeable clay of the basin. The reeds slowed down the water-flow and slowly the clay particles built up until terrestrial plants began to colonise the area. Eight thousand years ago, the Fen area supported oak, yew, birch and pine trees and was inhabited by deer, wild boars, horses and wild cattle. Three thousand years ago, the peat had reached a great depth, the waterways were full of fish and eel and there was an enormous variety of aquatic bird life.

The Romans made the first attempt at reclaiming the Fenland; their arrival coincided with a drop in sea level and they were able to improve the drainage of the area by cutting a series of straight canals. There was already a small population in the Fens, living mostly on the gravel "islands" of Haddenham and Ely and later on, in the period between 1100 and 1300, when the sea level dropped still further, this population rose rapidly.

The traditional inhabitants of this rich and unique ecosystem found in it everything they needed for survival: they used the underlying clay for making bricks and the reeds for thatching their roofs and for bedding for their domestic animals. They dug the peat for fuel and their cattle pastured on the few fields in the drained areas. They gathered wild edible plants and berries and hunted the game. They fished in the waterways and cultivated vegetables in that exceptionally fertile earth.

Here was a bioregion that supported myriad sustainable, interdependent lifecycles up until the moment when private enterprise intervened and upset the equilibrium.

The history of the Fens now passes into the hands of the Duke of Bedford. In 1630, the Duke set up an investment company, appropriately called The Adventurers, with the intention of draining the Fenland so as to use the fertile land for large-scale agriculture. Bedford called in a Dutch engineer called Cornelius Vermuyden, who drew up and carried out the plans. The wealthy investors in return for their capital received large acreages of the reclaimed land.

The new landowners were delighted at first with the excellent yields from the Fenland, but the birds and animals native to the Fens had lost a large part of their habitat. In the wake of this decrease in the wildlife, the traditional Fen dwellers found their livelihood impaired. Their protests, however, went unheeded.

At this point, a new phenomenon appeared: as the peat soil dried out, the level of the land began to fall. Desperate measures were necessary; the canals were excavated to a greater depth and dykes had to be constructed along the banks of the rivers. Windmills pumped water from the lower Fens to canals and rivers that were now at higher levels than the land they were supposed to drain. The level of the peat topsoil continued to sink. Whenever there was a high wind, the dry, lightweight peat blew off the ploughed land: to be caught in what the local people called "Fen blows" was like being out in a sandstorm.

Around 1830, the wind pumps were no longer able to cope with the enormous amount of water and the new steam pumps were introduced. In the last sixty years, electric and diesel drainage systems have been used.

A huge area of the Fens is now well below sea level and requires an expensive system of locks in order to keep the arable land drained. All the rivers have high dykes and some of the bridges on the roads are a good five meters above the level of the surrounding land. A seven-meter pole, completely buried in Holme Fen in 1848, was visible in 1930 sticking out 4 meters, despite the fact that drainage of that particular area had been abandoned in 1890.

Nowadays the Fens are farmed at enormous profit with a truly staggering use of chemicals. A doctor in the village of March, where my brother lived, explained to him, with an air of embarrassment, that during periods of heavy rainfall he always found his surgery full of people suffering from allergic reactions and vomiting. He was aware that these symptoms were linked to the pesticides sprayed onto the fields from aeroplanes but felt powerless to make his voice heard.

However, it seems as if the story of the Fens will end with a victory for nature. It is clear by now that the peat continues to dry out and the level of the land is still falling. In some areas, nothing but the underlying clay is left. When the agribusinessmen realise they are no longer making big money from the Fenland, or when the State decides that it is not economically worthwhile to spend such huge sums pumping water out of the Fens, the sea will reclaim the whole area and the villages will disappear beneath the waves.

An ecosystem which took 12,000 years to come into being and which supported a rich diversity of life might disappear entirely in the space of 400 years and we must remember that its destruction was not the work of the human inhabitants, but of human greed on the part of a small group of speculators.

So a bioregion is an area with specific characteristics which need to be respected for the good of all the inhabitants, human and non-human. Bioregionalism, in its effort to avoid future disasters of the kind that befell the Fens, places its hope in the work of those who live in and love a particular place. And the key word is *love!* Rational thinking is a respectable activity but it doesn't seem able, alone, to stop such catastrophes. Practical life needs heart and the heart doesn't speak in figures, but through the imagination.

Here is why the movement in America has a poet's voice - I'm thinking of Wendell Berry, Gary Snyder, Gary Lawless and others with their elders standing behind them - Thoreau, Whitman, Ginsberg, Muir. And on the other side of the Atlantic, people have read the works of Ted Hughes, Patrick Kavanagh, R.S.Thomas. These contemporary poets speak of their places from the heart, but they don't romanticise them because they were people who in their everyday lives had intimately to do with the physical place they inhabited.

The English antipastoral poet George Crabbe wrote in 1783, *"what son of verse could share/ the poet's rapture and the peasant's care?"* He was speaking to the poet Goldsmith, who idealised rural life and the happiness to be enjoyed in poverty. And this rejection of pastoral poetry comes down to our own day, because we see in this tradition a desire to shut out the hard reality which has often been the experience of farming people down the ages. What of the English guilt for the Irish famine in 1847? What of the forced removal of Scottish peasants during the Clearances of 1700-1800? What about the historical indifference of the Italian state for the landless peasants of the south, the men shot dead by the police in the village of Melissa in 1949 for daring to plough and sow unused land? Political power has often found it convenient to pour a gloss of romanticism over the lives of the poor.

But now we are faced with a new social phenomenon: there is a growing number of people in the industrialised world who have decided to go back to farming or small-scale craft for what we might call inner reasons. What are these reasons?

What induced Wendell Berry to abandon the academic world in favour of the family farm in Kentucky? Or led Gary Snyder to build his own cabin on the Sierra Nevada and live a life which keeps his hands dirty?

I think they both had a good look at what the world of privilege has to offer and what it asks in return: it asks us to give up the ancient, magic communion of hand and eye which engendered human culture; it asks us to give up the daily embrace of the physical world in its terrible beauty. And so we would have to turn our faces away from the violent, elemental energy that our real participation invokes, and from the feelings lived through slowly and heartily at walking pace. These poets speak to us of the physical world and their deep participation in its workings, and they speak without any gloss.

We have surrounded ourselves with apparent security but now we find ourselves without a place that recognises, comforts and sustains us. The power we have given away has come into the hands of small, distant groups who have turned our places into sources of profit. We have got rid of our shamans, and so no-one helps us mediate the inner and outer creative forces and we are afraid of opening ourselves to the healing wildness that can be predatory and destructive without the rituals for containing it.

Despite our educated, first-world mode, we are in a desperate situation, needing to reclaim our selves from the impersonal, ruthless powers who have taken control of our lives in return for convenience and easy living. The serf at least saw the face of his owner. How do we go about reclaiming our independence?

Ordinary people feel this need, and express it in unexpected ways. Last year I was in London: it was a hot June day and I was staying in Brixton, a live, black, exciting part of the city. Suddenly, the traffic stopped, there were banners everywhere: "Take back the streets!" Crowds of people sat in the middle of the road and ate picnics, there were people playing music here and there, other people dancing, Rastas singing, lads climbing up the street lights and looking down on the police, who talked worriedly into their mobile phones. I looked up at the banner they were hoisting across the main road. It said, "UNDER THE ASPHALT RUNS THE RIVER! DIG UP THE EFFRA!"

I thought Effra was a road - Effra Road. I had no idea the Effra was a river. I asked a boy. "Oh yes, the Effra runs right under our feet - they've buried it but it's still there!" he replied. I was moved to find people expressing their anguish at the inhuman dimension of the city with the image of the invisible river, as if the running water were a lost and hidden part of themselves.

The Welsh poet D.S. Thomas wrote:
"In cities that have outgrown their promise people
are becoming pilgrims
again, if not to this place
then to the recreation of it
in their own spirits."

(from *Later Poems)*

The idea that health flows towards people from the place that contains them is old, but to see a crowd of urbanites in the middle of Brixton turning in spirit to the invisible river under their feet was a real demonstration of how the natural world is the foundation of our psychic wellbeing as well as our physical life.

The day after this jubilant homage to the river Effra, I heard a story about another London river. Jill Purce, who teaches Mongolian overtone chanting, told us how she had offered to go and do some singing with the children in the Hampstead primary school where her own children go and had discovered that the source of the river Fleet actually rose in the basement of the Victorian building that housed the school. The Fleet is the river which gave its name to Fleet Street, the historical home of all the major British newspaper offices.

Who remembers that the Fleet was a river? It disappeared under the city streets years ago! Jill was curious and asked if she could go down to the basement and see for herself. There she found a large Victorian cellar with a vaulted ceiling and a stream of water pouring out of a rock in one side of the room, flowing through the cellar and away underground. The thought came to Jill that she could do a ceremony for the river with the children, none of whom had any idea that there was a secret, underground spring below their school.

While she had been down in the cellar, Jill had noticed a dusty old chair and a ball near the stream of water. She asked the headmistress to have them thrown away before the day she had chosen for the ceremony. The headmistress seemed a bit embarrassed and changed the subject. So Jill asked the deputy Head, then the school secretary and finally, the caretaker: they *all* changed the subject and avoided catching her eye. On the morning of the ceremony, the old chair and ball were still in place, so Jill asked her husband to throw the rubbish into a corner, out of the way. The children climbed down the stairs holding candles and each child threw a coin into the stream as a sort of offering and then they all sang a song for the river together, pleased to have an adventure instead of ordinary lessons.

Six months later, Jill had a sudden urge to revisit the Fleet, so she took the key to the school's cellar and went downstairs. When she opened the door, there was the chair with the ball, put back in precisely the same position as they had been when she first saw them! She went and found the headmistress: "Excuse me for asking," smiled Jill, "but what *is* that business with the chair and the ball down in the cellar? The headmistress blushed in confusion and finally explained that the chair and ball "had always been there" and that, when they had once been cleared away many years ago, a series of disasters had overtaken the school: the cook had got burnt, the exam results had been terrible, there had been a bad outbreak of whooping cough amongst the children. So it had seemed a good idea to put the chair and ball back by the flowing water, whereupon everything had taken a turn for the better.

The throne and sceptre of the river god? Somehow these twentieth century urban people who found themselves in the role of the four guardians had recognised the magic nature of the hidden source and understood it was a "hot place" requiring respectful behaviour.

After Jill had told this story, an Irish woman who had been listening asked if anyone had noticed the ugly cast iron tube covered in soot running obliquely above Sloane Square underground station in central London. "Someone told me that a river runs through it. They couldn't get rid of the water, so they imprisoned it in that tube."

Here is *The River,* the poem by Ted Hughes, poet and farmer: we went and pasted it up on the platform at Sloane Square station and at Effra Road, the next day, as an act of acknowledgement of these hidden rivers:

Fallen from heaven, lies across
The lap of his mother, broken by world.

But water will go on
Issuing from heaven

In dumbness uttering spirit brightness
Through its broken mouth.

Scattered in a million pieces and buried
Its dry tombs will split, at a sign in the sky,

At a rending of veils.
It will rise, in a time after times,

After swallowing death and the pit
It will return stainless

For the delivery of this world.
So the river is a god

Knee-deep among reeds, watching men,
Or hung by the heel down the door of a dam

It is a god, and inviolable.
Immortal. And will wash itself of all deaths.

Ted Hughes began working the land in Devon at the beginning of the 70s and some of his poetry was written around his experiences amongst the cattle and in the fields and woods of his farm. He said that when he first came to Devon, the Dumnoni people were still there, "the people of the deep valleys" the tribe which the Romans found living in that place, and he said it with respect and admiration for an ancient culture which had survived so many centuries. This culture of place collapsed during the nineteen seventies under the onslaught of agribusiness and Hughes watched it happen.

The vitality of the day to day work breathes through his writing, which embraces the wild world, the underworld, and the invisible in its sudden and sometimes violent material epiphanies. Some poets, said Hughes, "qualify for the magic drum" and he knew this frightening truth for our times: "the inner world, separated from the outer world, is a place of demons. The outer world, separated from the inner world, is a place of meaningless objects and machines...it is the IMAGINATION which embraces both outer and inner worlds in a creative spirit."

This is why the movement back to the land has been translated into poetic images. Poetry is a shamanic activity which binds us together in this new life, it is a practical way of talking to us about the physical world in its sacred vestments. Hughes said, "Shamanism is not a religion but a technique for moving in a state of ecstasy among the various spiritual realms and for generally dealing with souls and spirits in a practical way, in some practical crisis."

Certain poets, bards, storytellers have come forward to act as shaman for us. With gratitude we acknowledge their right to the magic drum! Snyder says "We tell stories to help preserve the world".

I spent my last night in London before I came back to Umbria at Crystal Palace where a large group of eco-warriors were camping, at the behest of the local people, to protest against the local government's plan to build a huge shopping mall and multistory carpark on that green hill. There was a camp fire and tea brewing (they might be eco-warriors but they were still English, after all!) and there was quiet talk about the vegetable gardens, and about the tree-houses and underground houses being built to hamper the bulldozers.

A guest turned up, sat down in the circle by the fire, thanked the people who were camping there for their stand against the mall and for putting up with the lack of home comforts and then offered to tell stories. He catalysed the group round the fire for an hour with his beautiful, ironic poetry, overflowing with respect for the world of the joyful and grimy-handed. Afterwards he told me that he never writes his poetry down, because then it would disappear: a real shaman.

The bioregional vision of a re-enchanted, reinhabited earth
"*is no paradisal dream.*
Its hardship is its possibility." (Wendell Berry, *A vision)*

January
Lilly's lamb

I climbed up the hill, with the afternoon turning colder and darker. The sheep had come down with the ten lambs that had been born so far and were hanging round the courtyard outside their stable as if they wanted to get indoors and it had seemed to me that some of them were missing. The wind was icy cold, I had working gloves on but I could feel the intense cold on my face. I had been several hours making bundles of firewood for next winter on the lower field, trying not to give in to a desire to go home. The field was covered with heaps of branches - oak, field maple, juniper and old vines, the perfect size for cooking wood. But it was a slow job and not much warmth in it. The three dogs had kept me company, in a real show of loyalty, but they were trembling as they tried to snuggle down in the dry leaves.

When I got to the top of the high field, I found a sheep with twin lambs already up on their feet. There were splashes of blood on the dry oak leaves underfoot and trails of jelly-like afterbirth, but the sheep had got round the back of a juniper bush now with the lambs, out of the wind. There was a male and a female: good, Matilda is a very good milker and we will raise her female lamb. Then I saw Lily, an old sheep, lying nearby, her fleece shaking with contractions and a lamb's head already out with the little hooves alongside its face. I wondered how long she'd been there. It was hard to tell if the lamb was dead or alive and I wiped the mucus off its nostrils and waited for the next push. Lily seemed very tired and the legs should have been further out. When the wave came, I pulled the legs out and the lamb slid halfway. Another push and it was born. It was moving - it was OK! I put the little body in front of Lily's mouth and she began licking at it but showed no signs of getting up. After a few minutes, the lamb found its feet and began to circle Lilly, staggering and rubbing itself up against her in the cold wind. I was sure Lilly had another lamb still to come.

Martin appeared and picked up the twin lambs to take them down, Matilda running alongside him, bleating. We decided to leave Lilly out for the moment because if she was going to have another lamb, it would be better out here than inside the crowded stable where she might mislay one of them and then reject it. She's a big sheep and she's always had twins but in the last two years she'd refused to acknowledge more then one. Last year, we managed to get the orphaned one adopted by another sheep but it hadn't been easy: it had taken three whole weeks of bottling feeding and persuasion.

I didn't want to leave Lilly alone so I moved a little way off so as not to bother her and sheltered from the wind behind the same juniper bush where the twins had stood: there was no wind there but the cold was still intense.

This is the twentieth lambing season for me. I remember the first time I saw a sheep give birth. It was Little Lamb, our first sheep. At the time, she was the only one we had. We had been peeling maize cobs at our neighbour Orlando's and he wanted to give us some maize in return for the help. "Orlando, if you wouldn't mind, what I'd really like is to get my sheep pregnant" I told him, "Could I bring her up to your ram?" "Course you can! Bring her up!" he said. So we shut Little Lamb, young and ladylike, in the shed with Orlando's huge ram. She wouldn't comply; the ram behaved in a courteous way for a couple of minutes and then, when she wouldn't heed his gentle, throaty respects, he slammed her up against the stable wall without further ceremony. Orlando and I were crouched at the window, watching. I felt like a Victorian mother who has sent her innocent daughter to her fate on her wedding night and even Orlando seemed upset by his ram's rough manners. "Oh, ssssh, beastie!" he protested through the window. Anyway, Little Lamb came home pregnant and had her first lamb that winter. Martin and I and Melissa, who was five then, carried out supper plates down to the stable and kept her company. We sat on the straw bales and I felt every contraction as if it were mine because this sheep was an orphan lamb and I'd bottle fed her. Her baby came out big because Orlando's ram was a large "nostrale" sheep and not Sardinian like her. She was the founder of our small flock of twenty. Little Lamb died years ago, but there are still daughters and grand-daughters of hers in the flock and Lilly is one of them.

After half an hour, Lilly struggled to her feet and I went and helped her. I picked up the lamb and we started down the hill. Once a sheep loses the smell of the lamb, she panics because they don't rely on sight as much as we do, so when you carry a lamb, you have to keep bending down and letting the mother smell it every few paces so she knows it's still there, but Lilly was an old sheep and she didn't worry. Martin had made her a separate place in the stable with bales and we put her there with the lamb. She no longer had contractions but she looked exhausted. A little later, the contractions started again and Martin spent half an hour with his hand inside her womb, trying to turn the next lamb the right way round. It had its back towards him and the legs folded under its body. In the end, he managed to pull it out, but it was dead. It's always a shame, but it was easier on Lilly, who couldn't have fed two.

When we had given hay and water to all the animals - sheep, donkeys, horses - and shut the hens up, we collapsed in front of the fire in the kitchen and we were so tired after that long, cold day, that we sat hypnotised by the flames. We made a bigger fire than usual and the pleasure of the heat and light from the fire was so intense that we sat without talking, our eyes on the hearth, our hands and feet thawing slowly. Camilla came home and - what a good daughter! - understood how we felt and set about making hot soup on the woodstove with onions, lentils and cabbage.

The hours spent out in the cold wind seemed to me to make our enjoyment of the fire fiercer, the pleasure of hearing the crackle of the twigs burning came partly from having handled those branches one by one last winter. That dead lamb rendered even more precious the tranquillity of the stable full of little new born lambs sleeping in a big heap together, just as the memory of days hoeing in the garden under last summer's sun drifted from the steaming plates of soup on the table.

February
The new wine

Martin brought up a bottle of the new wine today so as to offer a glass to a Canadian visitor, Alon, who's happened down the path in the middle of winter. It was a very cold day: soon enough it will be Carnival time and the wine brings the summer back to mind! We know it's good because we tasted some during the olive harvest, at the end of November.

That day there had been just the two of us: in the morning, we had already picked half a quintale of olives and we had such cold hands, we opened the barrel and drew off a litre of the new, untasted rosé to warm us up at lunch and lift our spirits. "I don't know what you did with this wine," I said to Martin that lunchtime, "not only is it delicious, but it makes you *laugh!"* And we had, in fact, drunk the whole litre, laughing like two madmen. When it came time to go back and pick, we looked at each other and Martin said "Do you think we're in a state to climb up olive trees?" "I really don't think it would be wise," I replied; and so instead of going back to the olive grove, we went to bed.

Alon drank deep and laughed a lot too and told us some stories from Quebec, where he comes from. One of his stories was about the time he came home from university to find he was shut out of his quartier because during the day a wall of barbed wire had been erected round the whole area to protect the VIPs who were meeting there to discuss global trade - presumably from people like Alon. The soldier who was guarding this unexpected wall advised Alon to go and get a resident's access permit for the area from the office. "Is the office open now?" asked Alon. The soldier had consulted his bullet-proof watch: it was five in the afternoon. "Tomorrow morning, nine o'clock," he said. "And where shall I sleep tonight?" asked Alon. This time the soldier didn't reply, merely fixing his eyes on a distant point, with an expression redolent of serious military secrets.

Alon was Jewish, and while we talked about wine, he told us about one of the Jewish traditional feasts, Purim, which comes round at the Full Moon of this season. He explained there was a lot of drinking and dancing and it seems it is important to get drunk at Purim because this is the moment of Chaos. People wear masks to hide their identity, just like we do at Carnival, so as to feel free to behave with licentious abandon. With the Purim story, there are two legendary figures involved: the hero Mordecai and the wicked Haman and the idea is to drink until the distinction between "blessed Mordecai" and "cursed Haman" is blurred: it is important to perceive that these two figures are only the two faces of the same man, a sophisticated concept. In our own tradition - both in the Saturnalia of ancient Rome and in the Christian Carnival - the period of license evokes Chaos. In our everyday lives, we're constantly trying to create order (usually we keep our noses to the grindstone all the daylight hours in the olive-picking season!) but we know in our hearts that Chaos is the Mother of Order and that, every so often, we have to fall back into the matrix and let go.

This thought must stand as explanation for the many empty wine bottles left by the fireplace this evening.

March
Sowing potatoes on the day Raja's foal was born

I was making the first cheese of the year when I heard Martin calling outside the dairy "Shall we plant the potatoes this morning - it seems nice weather?" It was March 17th and I thought "It's Saint Patrick's day, patron saint of Ireland, surely a good day for putting in potatoes!"

It was already a good day because at dawn we had found Raja's new foal lying on the hillside enjoying the first rays of the sun, with its mother nosing it gently and the other two horses watching from a respectful distance. The foal was already dry: it must have been born last night. The horses were on a very steep pasture so Martin opened the gate and led Papu and Imperia down, leaving Raja with her foal to make their way down to the stable at their own pace.

We started on the potato job. There was a young German woman called Miriam, who had hitched here the night before, helping. She hoed barefoot with her long, blond hair blowing in the spring wind. We were very pleased to have found red seed potatoes from Colfiorito instead of the usual Dutch potatoes. Potatoes for sowing are supposed to come from higher up so as to avoid diseases and the high plain of Colfiorito is 900 metres above sea level. In a separate patch, we sowed a hundred potatoes that a friend from the mountain village of Costacciaro, behind Gubbio, gave us: they are called "Campitello" after a tiny village in the Apennines which has its own local variety. It seems there were hardly any of these potatoes left and we wanted to help save them, so this year we planted them just for seed, to share with two or three friends for another year's sowing. It will be very interesting to taste them and anyway this is part of a general effort nowadays on the part of farmers to save their own seeds and prevent globalisation depriving people of their local varieties.

As we worked, the sky kept changing, fast clouds passing in front of the sun. Every so often, it got too hot and we took off our jumpers and Tshirts and then we'd get a shower of cool rain on our bare backs. Papu, the stallion, called endlessly from the nearby corral: he was upset because he couldn't see Raja anymore and, in the end, Martin went up to the pasture above the woods and led Raja down, with the foal running lightly alongside her. What long legs foals have! These horses are from the Camargue marshes of France. They are white but the foals are always born dark brown: it must be a protective camouflage and in fact, as Raja and the foal came down through the broom, the foal was far less visible than its mother.

While we sowed the potatoes into the trenches, we told Miriam what happened a few years back with the potato planting. That year, we had decided the plant them in a new garden, down by the stream Cerquetelle, and we had gone down with the donkey, laden with fifty kilos of seed potatoes. We put them all in and the next morning, we went down again with another thirty kilos. But when we arrived at the garden, what was this? The earth was all dug up and most of the potatoes were scattered across the surface. We realised that it must have been the wild boars, digging in under the fence and having themselves a feast in the night. It was a big disappointment: we had been looking forward to having a garden near the stream, but we had to gather the potatoes back up and load Otello again and go and sow them in the old garden near the house. I remember Otello's expression of resignation when he realised that having carried fifty kilos of potatoes down one day and thirty the next, he was now going to have to carry most of them back up the hill again. He must have felt like those poor mules in the Swiss army, who spend their lives carrying sacks of sand up and down the Alps in an endless simulation of wartime activity.

When all the potatoes were in, Martin and Miriam and I sat on the slope just above the corral and watched the new foal and it seemed like one of those rare days when you take two steps forward and *none* back!

April
Noah!

It had been raining for three weeks. It was just as well because it hadn't snowed at all that winter and we were beginning to worry about the summer, when we might be short of water. Round here the proverb says "Under the snow lies bread, under the rain lies famine" - but still, rain is better than nothing. And so we were grateful for the rainy days...but after so long squelching through the mud, we longed for the sun.

And then one evening, the sun broke through just before it set and we went to bed pleased: maybe the next day would be fine? Next morning, we looked out of the kitchen window and the sun was shining on a human figure lying in a sleeping bag on the field. Who could it be?

When he heard the sound of people moving about, the stranger got up and came to introduce himself. He was a young American. "Hi! I'm Noah!" he said. We looked at each other laughing. *Noah!* "So the flood is over!" we said. We made him some coffee and he explained that he'd got in late the previous night and hadn't wanted to wake anyone up. He turned out to be an art student from Vermont who had spent several months in London and was beginning to feel the need of some country life. When breakfast was over, Noah asked Martin if he had any work he could help out with "Yes, I'm pruning the olives at the moment and I'd be glad to have someone carry the olive branches down." said Martin. "Olive branches! That's my line of work exactly." joked Noah. Fortunately, he was twenty and not nine hundred and sixty-five like the Noah in the Old Testament, and so he spent a week dragging piles of branches to the woodpile. Sometimes I asked him to come and give me a hand in the vegetable garden, and he told me stories of his travels and his loves. He also talked about the land his father had bought in the Seventies where he himself hoped to go after his studies and combine farming with painting.

Noah was young and so, like all of us at that age, he had plenty to find out yet ("Who are the Etruscans?" he asked), but he was aware that Pratale was no rarity: the West is full of people who have decided to stop being customers and be as self-sufficient as they can. He had already stayed on various farms in the States and in England where people were trying to live simply and his own parents had gone back to the land in the Seventies. They had given up because they were overrun with deer hunters and his mother has narrowly missed getting a bullet in her head as she walked across the farm with a small baby in her arms.

So when I told Noah I was just about to go to a meeting of "new peasants" in Tuscany at the end of the month, he asked "A bioregional meeting?" In America, at the beginning of the Seventies, there was a movement of people back to the land and "bioregional" was a label a lot of them used. Unlike the national and state borders, drawn with rulers by the white invaders of the continent, a bioregion refers to an area, defined by the natural features of the landscape, the mountain ranges and the watershed, and the bioregionalists think of themselves as reinhabitants of a particular watershed and part of a specific natural community which includes not only humans, but also the plants and animals. They try to live as *part* of the web of life, holding an intention not to take too much in terms of habitat and resources.

Then a professor of sociology arrived here. He was a genial character, but very far removed in his views. He was "here" for two days but he never stopped travelling and in forty-eight hours, he visited most of Umbria. The last evening, he was home for supper and without any intention of giving offence, he remarked offhandedly, "Yes, of course, I know there *are* people who live this way - especially in Oregon. But it is an irrelevant phenomena." He obviously had no idea that the people at Pratale might feel not only part of their own local community but also feel part of this tribe, part of a movement of many thousands of people, nor did he realise that this tribe grows larger every day. I thought "No wonder the French government was able to make such a miscalculation when they sank the *Rainbow Warrior*. Or the authorities in Seattle when the no global movement blew up in their faces." Social phenomena don't start big but like everything else in nature, embryos have massive powers of growth!

I said goodbye to Noah and set off for the Maremma, where I went to see two friends, Francesca and Momo, in the country, surrounded by their horses and goats, chickens and pigeons. I had such a good time because I hardly ever go away and so when I do, everything seems so exotic! We walked on the sandy earth in the sunshine through the strawberry-trees - *corbezzoli,* the heather, ilex trees and cork oaks and they talked about their plans for the future. Their old house was full of tools for making saddles and bridles, home made ceramics, books on country living, herbs drying in bunches, gumboots, homemade cake and relaxed, smiling faces. I felt that the lack of money and the solitude didn't weigh on them and that they, too, felt part of the tribe who enjoys life more than most!

Francesca and I set off together in her old car - one of those cars in which travel is the adventure it should be because it *might* get you there or you might have to push or even abandon the car and hitch. But it took us all the way to Avalon, called after the magic Celtic island of apples which can only be reached if one is in the right frame of mind. At the fork in the track that led to Avalon, there was a rosebush in full bloom that seemed truly miraculous in April!

Here at Avalon live part of the Elfin People, a large, anarchic community who were hosting the meeting. They live without sanitation or electricity and without a lot of other things that most people find essential. We had two days of meetings under their olive trees or in the farmyard by the fire, sitting in a circle with a talking stick polished by years of use. The food was cooked on the fire and we had candles for night light. The Elves had their share of young runaways and old drunks to carry and carried them with a fair amount of tolerance and heart. I had to laugh, thinking about the image of the agricultural businessman that the Italian Farmers' Union would like to promote nowadays when I compared it to these men with their long hair and beards, their backs and arms tattooed with fairies and kitten's faces, their earrings and Sixties beads, and the women with wild hair and no make-up and fantastical, home-made clothes. There were sudden bouts of dancing and flute music at breakfast, constant games and intermittent drumming. But their vegetable gardens were weedless, their thousand olive trees were all pruned and hoed, the firewood was piled neatly along the paths, the Jersey cow was cared for and milked on time, the ducklings had a big, new pond to swim on. The Elves made pizza for sixty people in the wood-fired oven and stirred the warm milk for cheese-making over the fire. Up in the bedrooms, there were ordered piles of pages for the next newsletter. The human warmth of those two days and the strong feelings streaming between people as they embraced each other before going back to their own places did not look like an irrelevant phenomena!

May
Blackfly

Natural pest control: this is supposed to be a crafty human system for sending out hedgehogs against slugs, turkeys against vipers, spiders against flies, ladybirds against blackfly.

So a few weeks ago, I watched one of these ladybirds to see how she was doing. She was on one of my bean plants, there were blackfly beginning to take over and I wanted to see some serious action. The ladybird started up the beanstalk, turned off at a tendril and inspected a litter of little blackflies. She examined them carefully, one by one. But they didn't appeal, and she climbed slowly further up the stalk and got off at the next floor up. She crawled slowly over several leaves and looked closely at six different families of blackfly. I was getting impatient. Either this ladybird was not hungry or she was some sort of gourmet who wouldn't eat fast food, a ladybird for whom eating was an art.

She crawled right back down the beanstalk at an infuriatingly relaxed pace and started up another plant, where she went carefully over all the available blackfly before choosing one and eating it. Twenty minutes, this took. There were a hundred bean plants in the garden and at least fifteen had blackfly: what were the chances of pitting this ladybird against the enemy? I concluded that ladybirds are unreliable as pest control and when you think about it, what interest does a ladybird have in polishing them all off? Her concern is to have *plenty* of blackfly multiplying like anything so she can count on gourmet lunches and suppers for the whole season. The ladybird is not interested in beans.

Usually I reckon to share my vegetables with the rest of the biomass - what does it matter if we have ninety plants instead of a hundred? But last year, after twenty years of sharing, the blackflies had taken over completely and we'd had no beans at all, so now I was worried and on the defensive.

That year, I tried all sorts off nasty organic mixtures: fermented nettles, Marseille soap, tobacco delight, as recommended by various friends, all in vain. This year I decided to be the natural pest control *myself.* Every three days, I went over every single bean plant, killing the blackflies on the leaves, with or without offspring, one by one. I wish I didn't know that these aphids can reproduce not only sexually but also asexually and that they have dozens of babies every time, but I read this in the kid's biology textbook and this knowledge made me more ruthless.

While I went down the row killing blackfly, I began to realise that this whole business was more complex than I had thought. What were these ants doing, running up and down the beanstalks? They seemed to be doing something with the blackfly. I know that ants milk aphids, gathering the sap they have sucked from the plant, but it almost looked as if they herded them. Each litter of blackfly seemed to have a couple of very small ants standing guard, like baby-sitters - or shepherds? I knew there were ants who grow funghi in underground gardens, but have they also taken to raising domestic animals? Anyway, I killed all the blackfly I could find and went to have a siesta. That evening Maurizio stopped by: he's a naturalist so I asked him about the habits of ants and aphids. He confirmed my worst suspicions: "Yes, didn't you know? Ants raise the blackfly. You'll see - by tomorrow they'll have carried a new "seed" aphid onto each leaf!"

Maurizio was right! I did a whole field study and sure enough, the ants were so determined to raise their blackfly that they reacted to my forays by changing the position of the blackfly litters on the leaves. First they tried moving the litter to a different place on the upper leaf and then they moved them to the underside of the leaf, and finally, in desperation, to a point below the main vein of the leaf, so that is was impossible to squash the blackfly without damaging the leaf. These decisions were obviously taken at management level, because the changes in position were always simultaneous and identical on every plant.

The final result of the battle was quite alarming. I killed all the blackfly, but the bean plants refused to produce any beans. What did that mean? Either the plants couldn't stand the small amount of peripheral damage to the leaves or they were on the side of the blackfly: no blackfly, no beans! The ants are in the Neolithic with their cattle-raising and horticulture. What will we do if they reach the industrial revolution?

June
Toad in a tight spot

I pushed my way through the animals in the courtyard with difficulty: it was six on a summer evening and all the donkeys, sheep, turkeys, hens and ducks came out from the shade of the trees and surrounded me on all sides in the hope that I'd give them some buckets of clean, cool water instead of the warm water with bits of straw in it that were all they had left at that hour. Then they all stood in front of my feet, so that I couldn't move forward and the message was clear: "Shake the plum tree for us!" We have a lot of semi-wild plum trees here. The animals love them as much as we do and there's a sort of understanding that the trees in the old courtyard are "theirs".

The only relief for the eyes in this burning, dry season with the dust rising at every step, were the blue chicory flowers on the meadows and everything alive seemed intent on an endless search for cool and moisture.

That must have been why our bathroom was under siege every night. It's a room in the corner of the courtyard made of stone and wood and you have to go down two steep steps to get into it. I think probably all country bathrooms are ecological niches for scorpions, spiders and butterflies but the situation in our bathroom was becoming critical - out of control even.

If you visited the bathroom at night, you were almost bound to meet at least one toad, sitting there enjoying the humid atmosphere and no matter how many times we picked up that toad and put him carefully in the garden, the next night he'd be back in the bathroom- or possibly it was a new one. There is also a secret little hole in the floor where the taps are hidden that turn the electric pump over to the solar pump: this is the long-time home of a family of salamanders: every so often we lifted the wooden lid and watched their quiet family life and marvelled at their glossy black bodies and bright orange bellies and their new babies.

Last night I got up at four o'clock to go for what my old neighbour Rosa calls "making a drop of water" on the field and I saw Jupiter, Venus and Saturn rising close and luminous over the horizon. Then I went and checked the bathroom because one of my nieces said she met a hedgehog there last night. I find that hard to believe, but she must have seen something and surely she wouldn't mistake a toad for a hedgehog? Even though she lives in Hong Kong? No! She was right - there was a tiny hedgehog looking quietly up at me! When he'd had a good look, he crept under the hole behind the bidet. He must have fallen down the steps and then been unable to climb back out of the bathroom. There was the evidence - a series of desperate scrabble marks on the wall near the door. I put him down a saucer of milk and went back to bed. In the morning, the milk was gone and Martin was staring at the small pile of rubble under the door. The next evening we left him a plank to escape by.

Then this morning, we reached the apex of bathroom ecology. I heard a cry and when I went to look, there sat my little Chinese niece on the loo, waving a roll of toilet paper in her hand and laughing in astonishment. A very small toad was stuck firmly down the hole in the roll of toilet paper! He was trying to lower his head out of sight, but then his little legs appeared at the other end of the roll, pedalling frantically. How we all laughed! If it had been an adult who found him, they would very likely have screamed, but the little almond eyes shone with joy at this funny meeting.

Summer Solstice
Building the new house after the earthquake and dancing against the prospectors

The saffron-coloured silk cloth lay on the dusty ground of the building site surrounded by the circle of people and everyone took a turn placing something on it as a good-luck wish for the new house: a piece of bread, a piece of wood, a coin, a ball of spun sheep's wool. There was a picture of Saint Ubaldo, too, the patron saint of Gubbio and unrivalled protector against earthquakes, and a drawing of how we hoped the future house would look. We buried the silk bundle under the place where the threshold of the front door would be.

It was 15th May, an important day in Gubbio, because for many centuries this has been the day when the Feast of the Ceri is celebrated, at the very apex of springtime, when the oaks at last come into leaf, like the triumphant, concluding notes of a symphony: all the trees are now a luxuriant green and the bushes of yellow broom are flowering all over the hillsides.

There were twenty or so of us standing in a circle on the space where the courtyard of the new, wooden house would be. Jenni read the *Prayer for the Great family*, Snyder's version of a Mohawk invocation, very close in feeling to Francesco's *Canticle of the Creatures,* expressing gratitude to Mother Earth, to the Plants, to the Air, to the Wild Beings, to Water, to the Sun, to the Great Sky. It is not easy to make new rituals, but our carpenter friend, Benedict, is at home with this sort of process. Back home in England, he dances, in the country parts of Herefordshire where he lives, with the Morris side: six men and a violinist, who dance at fairs with their faces blackened, wearing ribbons and bells and brandishing sticks. Sometimes they have an accompanying Green Man. It is a very ancient tradition, and many of their dances are done on ritual occasions such as the blessing of the cider apple trees, when a libation of cider is poured onto the roots of the trees at the Winter Solstice, in gratitude for the fruit.

The ritual for the new house also ended with glasses of wine and good cheer, but in the middle of it all, a stranger turned up. He spoke Italian mixed with French. "I am from the Institut Geophysique de Lyons..." he began. "Have a glass of wine" offered one of the revellers. Perhaps it was the effects of the wine, the fact is that the newcomer, who seemed to be trying to announce the arrival of a seismological research team, and spoke a lot of technical French, let slip the name AGIP. "AH!" said someone, pausing in the act of pouring more wine and all the revellers stopped revelling and gathered round, suddenly looking sober. Agip is the Italian State petrol company. "Agip!" repeated one of the men, "It did seem pretty unlikely that the authorities would be so efficient as to organise a seismic study just after an earthquake! But what you're really intending to do here is a bit of prospecting for gas or minerals, right?"

By the end of the interview, it was clear that the stranger was the PR man for a team of prospectors and that he was deliberately confusing the issue by speaking French and using a lot of scientific jargon, so as to give the local farmers the impression that this was a study done in their interest. The reality was that the team would drill deep holes and explode dynamite down there so as to have a view of the geology 10.000 metres below the surface. They were following a line clear across Italy, from the Tyrrhenian Sea to the Adriatic coast. They were not interested in our earthquake, or the precarious, damaged houses, or the homeless population living in tents, or our precious water springs, which might disappear with their explosions. They were poised to drive through the steep woods with their bulldozers, making roads for their massive drilling machinery.

Benedict had arrived a couple of weeks before this, with a group of friends. Everyone called them "the van people" and they were here for three months, come to help start work on the Californian pole house. Benedict was a carpenter, Jenni and Andy were builders and Alison was an old friend of ours with a strong will and strong arms! We all looked at each other. Martin poured the PR man another glass of wine and then we warned him, "Don't take this personally, but we don't agree with either the content of your work or the way you go about it, so we'll do everything we can to stop you."

When he had left, we all felt pretty bad. "How come on the very day that we do the ceremony for the new house, these people with their appalling plans turn up?" I said. Jenni replied "I think it is because at the moment there's a group of people here who represent the very antithesis of Agip: we're here to act as an obstacle in their path! Even though we have the house to build, we'll try and given them a hard time in this valley."

Next morning, someone chalked on the side of the prefab "IF YOU WANT TO BE A GUERILLA, YOU HAVE TO GET UP EARLY" and sure enough, at seven o'clock we saw ten men appear on the edge of the woods on the opposite hillside, their shoulders slung with equipment of various sorts. They were marching down the hill, tying strips of orange plastic tape to the trees as they went. (How we dread that orange tape in the countryside! It means someone's intending to make a new road, or cut the woods, or pour cement). For ten days, those tireless men appeared every morning, retracing their steps across the valley, putting back the plastic tape as fast as we removed it. They reminded me of ants: they just kept on doggedly plotting their intended path of destruction over and over again, apparently unperturbed by the fact that the plastic tape never stayed put. Alison and Martin climbed up the opposite hillside with a tape recorder and a camera and tried to talk to them about the effects their bulldozing would have on the steep, wooded slope. The men were ill at ease, but their only reply was to avoid getting photographed or recorded. Later that morning, we heard one of them shout across the valley to their team leader "We can't get through! The greens are in our way!" We stared at each other. "What are the greens?" we wondered. It seems extraordinary now, but that was the first time any of us had heard the expression. It was 1985.

Then one morning, as we were heaving the poles into place for the framework of the house, we heard the roar of an engine on the other side of the hill. It was them! And their bulldozer. We watched in horror as the bulldozer advanced slowly down the hill, clearing a wide path through the scrub above the wood. When they couldn't turn easily, they simply ate further and further into the hillside, with earth and stones pouring down the slope. When they reached the beginning of the wood, they stopped. The mess was incredible: it had taken them three hours to make it and now it would take thirty years for the damage to repair itself, that's if the new path didn't get used for motorcycling or by hunters in the future.

My sister Jane managed to steal the plans which showed the red line that indicated where the prospecting team was headed next and we spent some time visiting other farmers further down the line, warning them about the operation, so that at least the PR man would have to tell the truth. Our neighbour, Franco, who was an ironic and humourous man, listened with raised eyebrows to the whole story when the PR man got to him. The man concluded "...so then we will do controlled explosions so as to create graphs of the geology at ten thousand metres!" At that point, Franco looked the man in the eye and said *"Ten thousand metres*? No way!" as if access to his land depended on the exact figure. But it was no use protesting, the team had access permits from the appropriate Ministry. However, they had been careless and failed to get permits for every piece of land and in fact they had no access permit for Pratale.

Meanwhile, more and more people were turning up to help build. We slept in tents and worked long hours, but there were May love affairs, plenty of children and babies to play with, the meals were full of good talk and good food and the weather was perfect - sunny but not too hot yet.

Our friend Richard had managed to kidnap the small drill for putting in fence-posts from a recalcitrant neighbour who had been holding this equipment hostage for a year. The whole neighbourhood had breathed a sigh of relief when he announced he'd snatched it back, and we had drilled the holes for the remaining poles and put the rescued drill back into local circulation. The poles were all up and the wall plate was finished. We were all admiring this first framework when someone came to say there were four huge machines belonging to the prospectors at the top of our hill, just about to come down our road. They were actually going to use one of our access roads to get into the valley! We thought fast and decided we had better block their advance by parking our only two vehicles, the old Landrover and Benedict's van, sideways across both the tracks that led down the hillside. Benedict and Martin went and manoeuvred the Landrover and the van onto the tracks and we saw the watching workmen get back out of their machines and confer. We went back to the building site and picked up our tools again.

In the middle of the U-shaped framework of the wooden house was a big space which would be the future courtyard and during the building years it was constantly full of half-naked children playing with the off-cuts of wood, brandishing hammers and spare nails. Jenni and Benedict's baby Rhiannon was the smallest, only five months old, and she hung from one of the house-beams in her little baby-sling, watching the whirl of activity. What a lot of fun those children had while we built the house! They were surrounded by piles of sand, stones, wood, tools of all sorts and a stream of different people. They built lots of funny little buildings of their own and when they got tired of that, they'd go off and talk to the donkey or climb trees. Every so often there'd be a shrill little voice "Look where I am!" The adults would look up and see one of the children balanced on a twig at the top of a tree, swaying in the wind. With heart in mouth, we'd reply "Wow! Well done!" No-one fell, no-one hurt themselves on the building tools: there must have been a winged angel of construction hovering over that site.

It was lunchtime. Silvana had made pasta with potatoes, a Neapolitan dish, for twenty-five people, cooking on a tiny stove inside the prefab. Living in the prefab was like being inside a guitar, the noise level and the resonance was incredible and just to raise it a little, the donkey Otello would occasionally come by, rest his head on the doorstep and bray through the open door. So we ate outside at a long table under the olive trees. Halfway through the *pasta con patate*, we heard a stern voice shouting "Whose is this jeep? Remove it immediately, it's blocking the road."

We went round the other side of the prefab and found a furious *carabiniere,* who was examining his muddy boots with an air of deep displeasure. "Is that jeep your property? You have obliged me to come away from my lunch. You have forced me to get out of the police car and walk! Look at these boots! Brand new! Covered in mud! Remove the jeep immediately, it is blocking the way of these workers!" Behind the policeman, we could see the man from the Institut Geophysique and behind him stood his team of ten workers, with their arms folded. Their belts were hung with with important instruments of *tecnologie.* The PR man was waiting, confident that these irritating people would now be dealt with. It was clear that he had called the police and that he expected them to lose no time in insisting on access for the machinery.

"I am really sorry about your boots and your lunch." I began. I could hear from his accent that the carabiniere came from the south of Italy. He calmed down a little and I tried to make him see things from our point of view, how we'd just had a serious earthquake, how frightened the children and the animals would be when these depth charges went off, so shortly after the terrifying tremors. That we feared for our badly damaged buildings and for the water sources if there were to be more underground disturbance. How the bulldozers would erode the topsoil on the steep hillsides. The *carabiniere* admitted that the prospectors had forgotten to get an access permit for this piece of land, "But, signora, "he added, in a conciliatory tone, "after all, *who is* AGIP? It's really the State! And who is the State? It's you and me!" Of course in theory he was right, but I thought how little this was true of the part of the country he came from and I replied, "No! You and I are *not* the State: you and I are the people across whose land, for centuries, these people, who represent the power and the money, have trampled! You and I are the ones who have always had to watch in silence!" I saw in his eyes that I had made contact and from that moment I could see his reasoning veer 180 degrees so that two minutes later he turned to the PR man and said "Right. So the Signora here will remove the jeep because she cannot obstruct your work. But, unfortunately you have NO ACCESS PERMIT here so you had better

ALL LEAVE STRAIGHT AWAY!" and he gestured in a way which left the PR man in no doubt at all as to his meaning. The team looked amazed at this unexpected turn of events and their macho poses faded.

We returned victorious to the *pasta con patate.*

The next day a very fat, well-dressed man appeared. His fingers were full of rings. He was no less a personage that the President of the Minerals Department of AGIP. He did not wish to come down our steep road, so he sent for me. When I reached the top of the hill, he addressed me. "Signora! You have cost me ten days work times ten workers. Huh! You live in a democracy so you can say what you think. But I want you to know that if we find anything useful here, we will *obliterate* this valley!" His eyes swept across the green valley as it lay glittering in the afternoon sunlight. *"Obliterate it!"* he repeated, in a satisfied tone of voice. Then he added, "You know, it wasn't like this when I was working in the Cameroon, Signora. There, the place was littered with villages and blacks - they can't read so there was no point notifying *them.* We just went straight through with the bulldozers and knocked down whatever stood in our way! So you can think yourself fortunate!"

"And how did you feel about doing that?" I asked. "I was just doing my job, Signora." he replied. And left.

Work on the wooden house progressed. The telephone poles were up and the pile of old railway sleepers had been sawed into planks for the outer cladding. The roof structure was under way. Everyone was pleased to be able to get on with the work without constantly listening out for the AGIP invaders. There was also the vegetable garden to water and hoe, the sheep to look after and milk, the cheese to make, the hens and ducks to feed, as well as large meals to prepare and human relationships to weave and repair and untangle.

Then, early one afternoon, the bulldozers suddenly appeared at the top of Luigi's hill. It turned out they had managed to get an access permit for that side of the hill. Benedict said "Look, we can't stop them but we can give them something to think about. I'll go and do them a sending-off dance. I won't do a cursing dance: we did that once with the Morris side, in front of a pub that wouldn't let us dance at their place and the next day, the pub burnt to the ground. We were a bit shocked at the effect! But I'll do a dance to tell them the land doesn't want them here!"

Benedict blackened his face in the traditional way with a burnt cork dipped in beer. Tall and imposing, with his black face and two sticks, he looked pretty scary. "These guys with the machinery won't understand what we're doing," he said, "but that doesn't matter".

We all walked slowly towards the huge machines as they rumbled down the hillside in the sun. Everybody had picked up pan-lids, sticks or spoons to beat the rhythm for the dance with. The drivers watched us coming: no doubt they thought we were going to sit down in front of the machines. But as we reached the point on the hill where the machines would have to turn a corner, everyone started beating time. Suddenly a terrifying black figure was whirling and leaping into the air, beating his sticks together in the direction of the oncoming machines. The machines came to a halt. The dance was impressive in the extreme, partly because it looked so ritualised, and partly because Benedict danced on one spot so that he looked self-contained and as if the energy which poured from him had a specific intent. He must have danced for five minutes.

The man driving the first machine understood exactly what was happening. He looked thoroughly alarmed and made gestures designed to ward off the evil eye.

The sky, which had been clear, began to get darker. With a truly amazing, vertical leap, Benedict stopped dancing. Everyone stood back to let the machinery proceed. A minute went by. Then the men, stunned and bewildered, turned the engines back on. But what was happening? A loud roll of thunder crashed right over our heads. The sky was very black all of a sudden. The machines moved forward and the first one tried to turn the corner on the freshly bulldozed path. Then a deafening downpour of rain started. Benedict stood silently by the side of the track with one arm pointing the way out of the valley. The thunder and lightning increased and the rain was turning the path into a torrent of mud. The maple trees on the field bent under the onslaught of the wind and rain, as if they too were pointing the way out of the valley. The giant wheels of the drilling machinery began to sink into the mud.

The AGIP men broke down before this evidence of black earth magic and they jumped off their machines and began running away, up the hill. With lowered heads, they ran past the company of soaking wet men, women and children, who were now all silently pointing the way out of the valley. Benedict turned to me with an amazed and delighted smile. "I've never danced to such cosmic effect!" he said.

After three months work on the new house under the hot sun, the van people set off home again, leaving a note on the back of an old envelope. *How to finish the new house*, it read. It seemed really easy: there were only four lines of instructions and we were forever grateful for that moral support and the van peoples' assurance as they left "The new house *will* get finished!" It took three more years work and nearly a thousand people helped us with the building:

when I think about that piece of common work now, it still feels like a more staggering outpouring of energy than the earthquake itself.

July
The snake

What a lot of death there is in this summer heat! The hot African wind from the south has been blowing for several weeks now and the drought is made worse by this merciless hot air that dries out the earth. Wherever the sheep and horses walk, a cloud of fine clay dust rises under their feet and catches in the throat. Yesterday I saw two male doves fighting in the courtyard and despite the small size of their struggling little bodies, the dust rose like smoke around them.

The burning sun shines day after day without a cloud on this season's arid, prickly vegetation with its thorns and spikey, hooked seeds. The meadows are full of battalions of Charlemagne's yellow thistles and hard, spring-like little grasshoppers with their dry, clicking bodies. The leaves on the plum trees are withering already and turning yellow. It feels as if the liquid element had hidden itself, concentrated down into rare points of distillation, in eggs, in ripening fruit, in the armoured bodies of insects
and there's a desperate fight for survival going on, which only a few will win. The moist, careless abundance of spring has gone and we are all food for each other.

In the shade of the mulberry tree, I was plaiting onions when I heard a blackbird's alarm call. You often hear that call at the approach of human beings and so I didn't think of going to investigate. But the blackbird went on calling for a good twenty minutes and in the end, the sound became like a newborn baby's cry - impossible to ignore. So I got up and followed the call until I caught sight of a blackbird in a thicket of blackberry bushes and plum trees at the back of the house and as I came closer, I was surprised that it didn't fly off.

I bent down and stared into the dark shadow of the thicket and when my eyes got used to the half light, I suddenly realised that two steps away from me there was a large snake with its head reared up, looking at me. He had found the blackbird's nest and he had pulled the three fledglings onto the ground and was eating them, with the desperate, protesting mother bird watching from a low branch. I thought how many weeks she had sat on the eggs and then fed the little birds. They were big, nearly fully fledged: in another week they'd have flown the nest. But there was nothing to be done, even if I had put those birds back in the nest, the snake had won. He knew the secret place now and he'd come back for them. For the snake, the little birds are a rich meal in a desert of dust. He was a big, gold and black Western Whip snake, not poisonous, but quite aggressive.

A minute went by while the snake and I looked at each other without moving. Then he slid out of sight. I went round the other side of the bushes to see if he was hiding close by or whether he'd gone right away, and while I examined the little pathways that run in and out of the undergrowth, I saw a toad so gigantic that for a moment I didn't know what I was looking at. The toad, frightened, moved faster and disappeared into the bushes in the direction of the bird's nest. What was this, some sort of meeting? The blackbird was still calling but I retreated to the courtyard: we are all at the centre of our own story and I felt I was treading on peoples' toes here.

The next morning, I saw the blackbird. She was already building a new nest, this time higher up, on a cherry tree. The hot south wind was still blowing.

18 August
Luigi's death

There's a line of people walking along the path to the big oak tree, each carrying something: a bottle of wine, a whole sheep's cheese, a loaf of bread, an apronful of warm tomatoes just picked from the garden. It's Luigi's lunch. Another year has gone by since his death and we are all sitting under the oak tree and sharing food and memories again.

This time my brother Tim and his wife Averil are both here too. It's unusual for them to be here in the height of summer with this great heat, but their presence makes the remembering more intimate and we talk about how Luigi's death came about. While Crystal, Briony and Miranda, their three daughters, swing on the rope that hangs from a branch of the oak, their blonde hair flying, we call to mind the strange sequence of events that led up to it.

At the time of his death, almost a whole year had gone by since Luigi's father had called him back to Sardinia. A letter was something unusual in that family, but his father had got one of his sons to write on his behalf. "You must come home, your brother needs your help: he has large herds of sheep and horses now and a good number of pigs as well." The very same brother who had been the cause, twenty years before, of Luigi's lonely departure for "the continent"! His father's summons was surely a gesture of love and care: he remembered Luigi as a boy, he hadn't seen him for many years, couldn't imagine him as a grown man with a life of his own, perhaps just wanted him back with the family. Maybe if Luigi hadn't lost his mother when he was only six, she would have been able to talk to his father and make him reconsider, up there in the mountain village. Here in the valley, all the neighbours tried to convince Luigi that you don't have to obey a father when you are in your forties, but Luigi too had only boyhood memories of his father, from the days when the head of the family made all the decisions and it seemed unthinkable to him not to do as the letter said. I could make no headway against his resignation. It must have hurt him so much to say goodbye to all the people he had lived alongside for twenty years, leave a small son, sell the hundred sheep he knew one by one, turn his back on the little dogs Vispo and Pastorella who had been at his heels all their lives.

He wrote sadly from Sardinia, said he was lonely and wanted to come back.

On 17th August of the year after he left, I had gone over to the other side of the valley on a visit. A good friend from Naples, Silvana, had problems with her vocal cords: it was serious for her because she is an actress and her voice is precious. There was a Tibetan monk staying in Valdichiascio and he was a doctor. He was the guest of a well-off northerner who had made available to the Rimpoche and his entourage the whole top floor of his large farmhouse for an extended period of time and by now there was a steady stream of patients from all over Italy coming to consult him. He was a quiet, smiling man, who had suffered several years imprisonment after the Chinese invasion of Tibet. Then he had had the chance to cure the sick son of one of his jailers and this man had allowed him to escape, although he had had to run down the mountainside with bullets flying around his heels.

We had asked for an appointment for Silvana, to see if Tibetan medicine could help her. The little monk had sent a message: "Come on Saturday morning."

We got into the old jeep and drove across the valley. There was very little traffic on the country tracks and when we met another car, we always slowed down. But the man in the *Fiat 500* we met at the fork in the road waved for us to stop. "Are you from Pratale?" he asked. "Yes" we replied. "Ah, I was coming to your place. I'm leaving my son with you because I can't handle him any more." Silvana and I looked at each other. What on earth did he mean? I looked at the stranger. He was about forty, thin and pale with a tired face. There were two children in the back of the Fiat 500: a girl about nine and a boy a little older. I couldn't bear to think how the boy must feel as his father declared him superfluous. His expression was one of indifference, as if he were not being discussed. I wished I could just say "Fine" straight away but I felt I had to behave like a serious person, talk about it, reason with the man, find out what this was all about. However, we had an appointment. "You go on down to the farm - we have to go now, we'll talk when we get back." I said.

We arrived at the Rimpoche's place and made our way through the ground floor with its trappings of an expensive Milanese lifestyle: deep carpets, stereo music, beautiful antique furniture, beautiful women. We went upstairs and suddenly we were in a completely different world: incense, Tibetan drapes, young monks in saffron robes with courteous smiles. It seemed that the Rimpoche only spoke Tibetan, but some of his novice monks spoke Italian and some of them English. They were very sorry, but the Rimpoche wasn't there. "But he sent us a message to come this morning" I said. The young monks were full of apologies but they were unable to explain his absence. "Come back tomorrow", they said, their shaved heads bobbing gently as they made little bows. It was a mystery.

Back home it was lunchtime. There were a lot of people and dozens of children, because it was high summer and we were in the middle of building the new wooden house. The prefab smelt of spaghetti with basil and tomato sauce and at the long table outside I could see the two new children and their wan-faced father. After lunch, we sat and drank coffee with a few close friends and talked to the man with the *Fiat 500.* He didn't want to talk. He said he had an appointment in Ancona and needed to leave quickly. He spoke as if it were absolutely normal to leave his "difficult" son with strangers. Rodolfo was a good-looking boy with brown eyes and he looked as if he were a survivor, as if the death of his mother five years previously had not crushed him. His sister looked sad and obedient. She didn't seem to be any trouble to her father. Neither of the children said a word, nor did they join the other children who were playing under the olive trees. We were all very ill at ease during this talk. The only thing we managed to gather from the father was that, since their mother's death, the children had been looked after by an aunt, who had now insisted that he take back responsibility for them. Only he couldn't cope with Rodolfo and so... We repeated that he couldn't just leave the boy here amongst people he didn't know without at least staying overnight himself to make sure Rodolfo had settled in. I wondered if I should insist on having at least the number of his driving license, because he seemed so reluctant to establish any sort of relationship with us. Supposing he never came back again? In the end we said Rodolfo was welcome to stay, as long as his father left us an address and agreed to stay the night and leave the next day.

I have to say that one day many years later, musing on that episode, I suddenly felt *sure* that the daughter was being sexually abused and I have cursed myself ever since for not focusing at the time on that possibility. She left the next morning with her father.

As soon as his father was gone, Rodolfo came to life. He quickly made friends with the other kids and spent the whole morning wandering around the building site, watching the carpentry work. Then he found paper and coloured pencils and began a large, detailed picture of the new house and all the people working on it. The smaller children crowded round him to watch. He looked happy enough and I left with Silvana for the Rimpoche's again.

The Tibetan doctor was at home this time. The young monks smiled and invited us into the room where the Rimpoche was sitting on the floor on a cushion. It was the first time we had met. He was also full of kindly smiles and in a soft voice he asked one of the monks to translate Silvana's words. The murmur of a language so foreign was soporific, I knew I wouldn't be able to catch a single word of Tibetan and so I let my mind wander and the whole conversation -Silvana's explanations, the novice monk's translation and the Rimpoche's quiet words of advice - went over my head. Rays of sunlight streamed through the window and the curls of smoke from the incense sticks rose slowly so that the two worlds, the countryside of Valdichiascio and the Tibetan heights merged together. Suddenly the conversation was at an end: the Rimpoche gave Silvana some herbal medicine he prepared himself from Tibetan flowers and we got up to go.

But the Rimpoche said something to the young monk, who turned to me and said "The Rimpoche asks you please to stay: he would like to say something to you." I couldn't understand why. We sat down again and the Rimpoche turned to me and began to speak. He spoke at length, for perhaps twenty minutes. The murmur of the Tibetan words sounded calm, almost comforting , now. After several minutes, he paused for the translator.

"Every being has his path in this world, " the young monk translated to me, "We should be compassionate, but we cannot take on the destiny of another person. That destiny is his and even when it seems cruel, he will know how to embrace it with serenity. You should not burden your heart over the disappearance of another being, because it is his time for leaving. You should not feel in any way guilty or troubled: you should continue your own path in mindfulness, walking forward quietly." I was so surprised at these unexpected words, so disorientated, that much of what he said flowed away from me like water, leaving no trace, but the essence of his discourse was this: he wanted to offer comfort, support, a sense of tranquillity.

When the Rimpoche had finished speaking, I thanked him and got up, still mystified. It was as if he had wished to reply to a very specific question of mine which I hadn't actually voiced.

When we got home, Silvana told everyone what the doctor had said. We described the house, with its curious split in atmosphere and decor between the upper and the lower floors, the Rimpoche himself, the group of young monks who attended him but we didn't talk about his final speech. because it was too strange and we couldn't somehow fit it into our account of the morning.

It was siesta time. The "African" squad as the Neapolitans called themselves (because they prefered to get up late and work in the heat of noon) sat under an olive and played the guitar, and the others slept. Even Rodolfo and the other children were lying in the shade. I stayed where I was, on the sofa in the prefab, meaning to go and sleep but not going. After ten minutes I said to myself "Why don't I go and sleep? What am I waiting for?"

"Where's Etain?" I heard someone call outside the prefab. I looked out. It was Ivo, who lives up at the top of the hill. He came towards me, his face white, and said "Etain, I came to tell you that Luigi died this morning. Rosa told me and she told me to come down and tell you."

I sat in the grass underneath the cherry trees for a long time, asking him to repeat what he had said over and over again because I couldn't take it in. It was many hours before I could hear the explanations, the how and why of this death. What broke my heart was the thought of the many months of solitude that he must have suffered before he died; he was someone whose soul showed through, someone who lived very close to the place where the worlds meet. If he no longer wanted to be here, it would have been easy for him to leave. For his family in the mountain village, his death was inexplicable, a pain they could not have foreseen.

Rodolfo drew a picture of an enormous falcon, the biggest ever seen round here: it flew in slow circles over the farm that afternoon, right over the hillside where I had gone to sit alone. Of the many people who were at Pratale at that moment, only a handful knew Luigi and that was very painful. Only Rodolfo and I saw the falcon. I suppose that was because of the nameless convergence between our stories that seemed to have occurred in those two days.

I looked out across the valley and I thought "Who will carry this place in their heart now? I should have paid more attention, I don't know enough." Luigi knew every movement and contour of this landscape as if it were part of himself, every change in the light, the plant life, the wild animals, there was not a path or a wood or a field which was not entirely familiar to him. When the master leaves, the disciple never feels he has learnt enough.

September
This is Saint Peter's House!

Last Thursday was such a boiling hot day that when I had walked up to the main road, I was mighty glad to hear Signora Emilia, kind and welcoming as ever, call "Come up and have a drink!" Emilia is the blacksmith's wife; her husband is a man who can knock you up a donkey cart or any spare part you like in no time at all, but he is a man of few words. So Emilia does the talking for two, and over the years she has told me lots of stories. Today she tells me about the time, in 1944, when the English artillery were up here at Mengara, bombing the mountain above Gubbio where the Germans were holding Italian hostages in the basilica before the retreat. When she's finished describing the scene, she says. "Now look at all these ants in the kitchen! They drive me mad! Worse than the English! What can I do to get rid of them?"

Emilia asks me this because she has a sort of idea that I'm a bit of a witch. I ought to say "I don't know" but I can't resist fulfilling her fantasies, so I say "You get a piece of paper and you write '*This is Saint Peter's house*'. Then you prop it up at the eye level of those ants so they can read it and they'll go away" This religious method for getting rid of ants was taught to me years ago by a Roman friend.

How did I come by this reputation here in the country? I think it was because, when I first got here from Rome, I didn't understand that magic was still a real force to be reckoned with in the countryside, and so I slipped into the role unwittingly. People saw a single woman who didn't seem afraid to live alone in an isolated place, and that is already a good sign of a witch here, where the country women always call a woman neighbour in to sleep with them if their husband is away. (Is this for the sake of respectability or out of real fear? The lack of either is socially suspicious.) Then there were two facts which lent weight to the idea: I said I had worked in a pharmaceutical company and that my father was an astrologer. I tried explaining that I had only been the managing director's secretary but that didn't stop people from consulting me on their sciatica or backache. And it was only later that I realised what they mean round here by "l'astrologo": he is the magician you call in when the vet can't work out why the sheep are dying or when the ordinary doctor can't seem to come up with the right ointment for a persistent stiff neck.

During my first summer here, the neighbours took over their new farm from the previous tenants. They had invited us to lunch under the trees outside their house. Pina was a culinary queen and my sister and I and various friends had eaten our way through home-made tagliatelle with goose sauce, followed by sausages and chicken with the strong red wine from their vineyard. At the end of this long and relaxed meal, we were all sitting back in a state of grace, talking of this and that when one of the sons came running up, shouting "The cow is dying! She must have got into the storeroom - she's eaten almost a whole sack of flour! And I can't find her calf anywhere!" We all leapt to our feet and ran after him into the stable. The cow was lying on the ground with the whites of her eyes showing, her belly hugely swollen.

Pina took the cow by her wide horns and shook her head to and fro. "Cow! Cow! Get up!" she yelled. Her husband Franco took the tractor and went to telephone the vet and all the children stood round the cow, horrified. It was their first cow, and she had just given birth to their first calf and now she lay there as if she were dying. My sister Jane squatted down next to the cow and began massaging her belly. "She must be all bunged up inside. Help me massage her, maybe it will move things" she said. One of the sons went to look for the calf. Maybe the cow would feel more urge to fight for life if she had the calf near her? I knelt down. It wasn't easy to identify with a cow's belly: close up, it was so enormous, but I tried to imagine how she must be feeling. The cow's skin was hot and hard under our hands.

An hour went by. Jane and I had got into a rhythm. Pina had tried pouring grappa down the cow's throat, a technique she said had once worked wonderfully with a billy goat of her aunt's. One of the sons had found the calf and brought it into the stable and pushed it close to its mother. The vet arrived. Everyone looked at him anxiously. He observed the poor, failing cow with unsympathetic eyes and then turned to Franco: "You'd better kill it, otherwise you'll lose the meat as well." A common medical attitude! When he had gone, I heard Pina and Franco wondering whether they should call in another vet. The cow's life wasn't a question of money for them, it was their first crisis in their new home and they didn't want to give up on her.

The calf nudged against his mother and for a moment she opened her eyes. We redoubled our efforts on her belly. We massaged her until sunset and by then, the cow seemed to be making an effort to get on her feet. We set off for home, leaving Franco's sons to carry on the work. Next morning, very early, Franco was under our windows. "The cow SHAT!" he called up in his Roman accent, "She's perfectly alright now! I called that vet back to come and see her. I said to him "Here's the animal you wanted to butcher yesterday!" And he said "What did you give her?" I said "We massaged her belly for ten hours!" And you know what he said to that? He said, "Oh well, if you want to go massaging an *animal*...!" Franco smiled "You two - you're *witches!* Much better than the vet!"

It was not long after the cow episode that my mother and father came here on holiday. There was nothing here yet so it was pretty much an indoor camping experience. My father spent his time building bookshelves for me and while he went about this project, he found a way of communicating with Luigi. He spoke no Italian, but he found that if he spoke Latin, he could make himself understood perfectly. "It's just like Sardinian!" explained Luigi. My mother had a whole adventure while I was away in Rome during the week. One morning, she woke to find two huge sows wandering about just in front of the house. My mother, although she was a Londoner, had spent enough time in the deep country to feel confident about this, so she got herself a good stick and set off to take these sows up to the nearest neighbour. She drove them across three fields and managed, with a certain amount of difficulty, to shut them in an empty stable. Then she climbed the hill to Franco's house and, pointing downhill to his stables, announced in Italian, "Two porky ladies" while Franco laughed his head off and tried to understand what on earth she meant. Finally he understood, and replied "But I don't keep pigs!" And sure enough, those pigs belonged to Francesco at the castle and in the evening he came down and fetched them home, passing by our house to thank my mother for attempting to bring them back.

And so everyone in the valley heard that the *astrologo* was here and people started turning up to be cured. My father was taken aback but did his best using the ten basic homeopathic remedies he always travelled with. He spoke slow Latin as he questioned his unexpected patients and none of us realised at the time what an awe-inspiring effect this added to the homeopathy. My father grew steadily more amazed as the trickle of patients swelled but the apex of his fame seemed to have arrived when my mother opened the door one morning to a woman with a string of ten children following her. It was early and father was still in bed. The children filed silently into the room and stood at the foot of his bed and looked at him with big eyes. My father looked at me and started laughing. "What is the matter?" he asked. I translated for him. "The woman says that this little girl has white spots on her belly" The little girl in the middle of the row, who looked about eight, pulled up her jumper to show us. "How long has she had them?" asked the *astrologo*. "Oh, since she was born!" replied her mother. My father gave her some homeopathic pills, choosing them from his restricted pharmacopea, and off they all went.

A week later, the little girl's mother reappeared. "The spots have gone!" she said, opening her bag and filling our kitchen table with her home-made biscuits. That evening, my father laughed until he cried and said "It's just as well we're leaving next week, or I'd really end up playing Merlin here!"

Yes, it's the successes that are the real problem. A few days ago, I passed Emilia's house again, and she called out "Signora Etain! The ants! They've *all gone*! Do you think you could make it so my daughter finds a job?"

October
Nanao walks by

We were just picking the last few bunches of strawberry grapes at the end of the *vendemmia* when someone called Luca turned up and helped us as we searched through the vine leaves in the fading light. He brought us an amazing, elongated pumpkin that looked like a swan. We put it on top of the piano: it seemed a shame to eat it. When all the grapes were crushed and in the vat, we went and made supper and Luca told us how he was finishing a degree in sociology at the university of Urbino and wanted to do his thesis on the population dynamics in the Italian countryside. This was his "head trip" but it turned out he had his feet on the ground too. He talked about the olive trees in Tuscany where his family has a piece of land. He spent his childhood there and it was clear from his story how that was an important part of his past experience, and his future too because he intended to go and farm there once he'd finished college: I counted another victory for the earth against the global market: Luca will not be one more sociologist at the service of the multinational mill. He decided to sleep the night here and go on the next morning for Urbino and so we had an evening of talk. Luca asked us a lot of questions about our contacts with other people in the country, about the bioregional movement in Italy and about the two poets who are coming in a few days time to read some of their poetry: he had hoped to find them here already but he'll try and catch up with them next week at the Elves' community.

He set off and I thought about his visit. More and more people in their twenties come through here with plans like Luca's, people who could easily have a career but who have decided against that world. I understand that when they come here with their questions, what they are really trying to ask is "Are you happy with this way of life, after twenty years of it?" They want to know if we have wearied of it, if we've lost the vision, if we long for a more comfortable life, if we feel isolated, if we regret the choice we made now that we are older. If we are still really alive!

I hope that our silent reply to the unasked question is clear: that we feel enormously blessed with good things - trees we sowed ourselves, that are now tall, cover us with shade and fruit, animals whose grandparents were born here share our days, friends who shared our work fifteen years ago come back and their children, who rolled down our hills as babies, come back with rucksacks and chop wood and tell us their stories. The reply is that this life has everything worth having: beauty around us, real work and real food and good friends to share it with.

But this October, we had the chance to ask the same silent question of a real Dharma bum, a man who's kept his feet on the ground for over fifty years, half a century of travelling on foot around the world. Nanao Sakaki is part of the radical Japanese subculture, a maestro of life at the edges. The poet Gary Snyder, who with Allen Ginsberg met Nanao in 1963 under the bridges of Tokyo, said "Nanao's poems are not written by hand or head but with his feet". So this was the man who came here this autumn with the Maine poet Gary Lawless, who'd already spoken his wild words in our courtyard in a whirlwind of falling, yellow mulberry leaves two years before ("everything moving, everything/moving to/come together, come together and/fall apart again./the water running/the heart beating.") So that afternoon, we and all our country friends from round about were able to ask Nanao a silent question: "How does it feel to be in your seventies, owning little and wanting less and to have in your heart and body the memory of many nights of sleeping on the ground, many meetings with bears and wolves in the mountains of planet earth?"

Just as Gary and Nanao were about to begin reading, a huge downpour started. Fifty people tried to squeeze into the kitchen. Inside it was impossible for anyone to put both feet to the floor, outside in the courtyard we'd get soaked. "Hey, Nanao, where shall we do this? Inside or outside?" I asked., "OUTSIDE!" replied Nanao, laughing and opening his arms wide, water dripping down his face. So we made a circle of umbrellas round Gary, who started to read and it was good to see how many people were happy to risk getting soaked for the sake of Gary's luminous vision, his sense of how the rest of the natural world speaks to us:

"The iceberg has come
to speak with Nanao.
She is just beyond the window,
waiting beyond the light.
She has come a long way.
She has a message for us.
She is very shy.
If we look directly at her
she begins to melt away..."

When Nanao Sakaki read, in his slow, emphatic voice, it felt like seeing the tracks he left through the woods, the traces of a life where there has been time for everything, where the limits of personhood have fallen away. He identifies bodily with the whole world...when Martin went down to the cellar with him, where the new wine-must was fermenting in the barrel, Nanao came back dancing "I'm bubbling!" The next day, he took a solitary walk in the woods and said afterwards, "It's sad the way they cut the woods in Italy, but the people here are so joyful that they'll *see* that in the end" And Nanao's reply to the silent question was clear:

"So joyous
You forget yourself
You forget a serious appointment

You become a piece of firewood
Warming up
Flaming up
Singing up
Dancing up
You become ash".

How to make the joy flow? "Keep your feet muddy"!

In the evening, Giuseppe Moretti , who had organised the readings with Gary and Nanao, set off back to his farm on the Po river to see how the levees on the river were holding up against the rising waters. Now his farmland and his wooden house lie eight metres under water because the levees have been deliberately broken so as to save the towns downstream from flooding. But when he writes today, he talks about the Po with the same respect and love as ever. "I am not angry with the river, my neighbours in the flood-valley don't resent the water. The river is just following his nature, doing his job and, in my view, he has given us all a big lesson in bioregionalism, if only we would understand." *This* is the spirit that comes from having muddy feet.

November
Rosa's flowers, Baldo's death

I found Rosa in her garden admiring her huge white and yellow chrysanthemums. "I have to make do with these," she explained, "because He's never let me grow flowers!" I knew this was one of the few sore points in a marriage that's lasted sixty years. The economics of peasant life were close-cut: there might not be enough water for the fields and the animals and it couldn't be wasted on the luxurious beauty of flowers. Antonio didn't work his fields anymore and he no longer kept cows, but the flower question had become a family taboo and Rosa couldn't break out of that cage now. But chrysanthemums were allowed: they were needed for the graves of dead family members and if they were not home grown, they would have to be bought at high prices outside the cemetery. I gazed at them too: they were so beautiful with their enormous heads. They reminded me of the autumnal vases of my childhood because in England we decorate our tables with chrysanthemums: they are not reserved for the dead as they are here.

These chrysanthemums also reminded me of the death of another old man, the grandfather of a family in the valley. His name was Baldo and he was a sweet-natured, easy-going person, over eighty when I first knew him. He pottered about the farmyard on small errands and spent a lot of time sitting quietly under the enormous mulberry trees whose branches shaded the pig-styes.

I had gone over to Baldo's farm to fetch a cylinder of gas, but the old man said they'd given up keeping a store of cylinders. He knew that I only drove my jeep on the dirt tracks, for bureaucratic reasons, so he added, "I'll be going down to the village soon to get a cylinder for us, so I could get one for you at the same time." Baldo had an old truck, a tiny, three-wheeled *Piaggio.* "Come back Saturday!" he said.

Saturday I went back and found Baldo picking olives from one of the two trees near the farmhouse. "Oh dear, I'm sorry," he called, "I haven't been to fetch that gas yet. I just wanted to finish picking these olives - they're the last ones. I'm sorry - come and have a glass of wine! The sun is really hot today!"
My children ran off to play with Baldo's little six-year old granddaughter Cristina and I sat down with Baldo in the kitchen of their post-earthquake prefab. He poured me a glass of wine and we talked about olive oil and the merits of the different local oil-presses. Through the open door, I saw Cristina picking flowers in the garden and I heard her father yell for her not to pick the chrysanthemums. It was too late - the little girl had already picked a whole bunch of those death flowers and she brought them in to me, smiling. She was too little to know their meaning: they were her present to me and I thanked her. The flowers lay on the table between Baldo and me while he talked about times past in Valdichiascio and how sad it was now for the old people. "There were lots of people in the old days", he remembered, "There was no lack of company for the old ones: they used to sit in the sun and chat and pass the time."

In the end, it got late and I got up to go. I called the children and said goodbye to Baldo. He apologised again, "I'm sorry about the gas - I'll be going soon." "There's no hurry, Baldo, it's very nice of you to go" I replied. As I climbed into the jeep, he came to the window and asked, "Have you got onions? Wait a minute!" He disappeared into the barn and came back with a long string of shallot onions - a gift. We set off and Baldo stood under his mulberry tree with a smile and a hand raised in farewell.

When we got home, I parked the jeep and then I stood there, with the bunch of chrysanthemums in my hand. I felt uneasy, and foolish about my uneasiness. I didn't want to bring them into our prefab. I laid them on the ground, under an olive tree...maybe the sheep would eat them. I hoped the children wouldn't ask me why I did that because I didn't want to talk about it.

Monday I had to go to town. When we got up to the main road, there was a death notice on the wall opposite the end of the track. "Baldo Papafava" it said. I read it ten times, I could not take it in. A neighbour woman came down from the house opposite and told me that Baldo had gone into the village with his little truck and had been hit by a lorry as he did a U-turn. He was dead by the time he got to hospital. The funeral was that afternoon.

I went to his funeral. He was greatly loved and the big church of Saint Francis in Gubbio was filled to overflowing: I was amazed at how many people knew him. We got home late after a very long day in town and as I got out of the jeep, I noticed a scattering of chrysanthemum petals lying on the grass. I felt a stab of shock - so Death *had* come, sure enough.

The next day it rained and I spent all day washing clothes. It was a meditative job and plenty of thoughts ran through my mind. First I thought "I've killed Baldo! He had gone to fetch the gas, if only he hadn't gone..." Then, slowly, I began to calm down. Death comes at the right time and doesn't depend on casual incidents. I remembered how I had sat in front of Baldo with the bunch of chrysanthemums and how he had said "Sorry, I know I have to go, but I just wanted to finish picking....." and I had replied "There's no hurry, but when you go..." I had played the part of Death! Like a reminder to the soul of the living: "It's time to go now". In the end, I thought how many roles we find ourselves playing in a lifetime. Sometimes we play the part of Love or the Fool and sometimes the mantel we find on our shoulders is much darker. None of us is so important as to decide the roles: we take them on briefly, that's all. It is only in the slower-paced life that we live in the country that I have felt rising into consciousness these connections that are embedded and hidden in day-to-day life.

What about that string of onions? Many months after this, I came across a description of the symbolism of onions: tradition has it they are a magical protection against the baneful influence of women.

When we built the new house after the earthquake, we chose a mulberry tree to plant in the middle of the courtyard because it was Baldo's tree.

December
Eating Naomi

"How disgusting! I don't want to hear about it!" said our guest from Vienna at lunch on Christmas day. We were eating an exquisite stew made with the meat of an old sheep and the red potatoes that come from the village of Campitelli and lots of paprika and I had just said to her "This was a black sheep, she was called Naomi, after Naomi Campbell"

Years ago, when we were beginners in the country and very poor, we often used to count how much of the food on our table for lunch or supper we had produced ourselves. "This is our own salad, our own olive oil, our own eggs...maybe one day we'll have our own wine too..." It was like telling a rosary, weighing up how far we'd moved away from dependence on industrial food. In those days, we hardly ever ate any meat, and then gradually, the times we'd have a rabbit or a chicken became a little less rare. But it was always an occasion: we never ate another being without a big sense of gratitude, without deep appreciation, expressed in words as well as in bodily flourishing. And we never ate anyone we didn't know - it mattered that we knew this animal had lived a life moving freely in the sun with others of his own kind.

So that's why family and old friends joined in with remembering Naomi and tried to make our Viennese friend see that eating someone you know is better than eating someone you don't. Why hide from yourself the nature of your food?

"Yes, Naomi was something else! She used to stand on her back legs and eat apples off the trees - in fact, she'd hold the branches down and let the rest of the flock eat their fill of leaves and fruit!" "Mmm, she was quite a male sort of sheep. Every so often, when I was milking, she'd come up to me with her head lowered and give me a knock on the forehead, just a tap, you know!" "She was a present from Tonino - a good-luck, black lamb - a gift to make the peace between our two families." "When she was old, she had a bad time with lambing and her milk would fail sometimes... Now we can raise one of the lambs from last spring: there's a limit to how much pasture we have and life is a cycle."

Our food is many-layered: in the pecorino cheese we can taste winter days spent mending fences, summer mornings clearing the brambles from the pastures, nights lambing or looking after a failing sheep, small children begging to plunge their hands into the hot whey, dawn milkings. The red potatoes remind us of summer days hoeing and of the little village in the mountains near us where this local variety of potato has been carefully handed down for centuries. The olive oil smells of cold, sunny, November days up the olive trees, with the hard green leaves in our faces and our bodies balanced precariously on the branches. The wine tastes of grapes but calls to mind drunken wasps, mint crushed underfoot in the vineyard, young willow boughs twisted in our hands to tie the vine-shoots in March.

In the last few years, the stories that come with our food have multiplied because now we meet up and exchange what we produce with other small farmers around Gubbio: we eat dried tomatoes and they brings to mind Silvio and the strong flavours of his native Puglie, the butter reminds us of Harry's Bavarian cows and his dramatic arrival in Umbria one winter day many years ago, when the lorry with all his household goods turned over in the snow. The chestnuts smell of the abandoned chestnut grove in the Maremma marshes that Francesca cleared. The first time she brought baskets of chestnuts, she also brought the story she'd written about clearing that grove and I thought "Suppose the supermarkets were obliged to write down for us *in what spirit* the food on sale had been grown or raised? Very likely we would all stop using supermarkets in favour of real food, steeped in story."

Soon it will be New Year's Eve and we'll eat the pig's trotter given us by Gianni, whose baritone voice surrounds all his work with song, whether he's building stone walls or feeding the pigs. We'll cook the pork with the lentils grown by Maria Rosa, a red-headed friend who, when she was about to give birth, walked down from her mountain farm with her companion and - since there was no bridge then - stripped off and waded through the swollen February river with the icy water streaming over her belly. This is our recipe for the good-luck supper, food made of Gianni's singing and Maria Rosa's courage as well as the usual ingredients: a delicious start to the year.

Winter Solstice
Meeting Kidhr

Someone who turns his back on consumer society and tries to find a way of living in the wild world again is like a man who comes home after a long journey. He looks at all the old familiar places and wonders exactly what it is about that landscape that gives him this sense of belonging.

The twentieth century has tortured and vivisected and exploited nature, demanding replies to man-made questions and the more information we have squeezed out of her, the further away from us the most important knowing seems to have retreated. Does the world include us? Are we wanted here? Is there anyone out there who wants to talk to us?

Over the years since I have come to live in the country, I have had the sense of waking very slowly from a deep sleep, gradually opening my eyes and ears to what the physical world has to say. Half-way through a piece of work on the top field in the early morning, suddenly I look up and watch how the mist rises above the red and gold woods, like a slow sea-tide. I stand in empty contemplation. Emile Mâle says that everything we now see as beauty, for the Celts was religion: the forest was divine, the looming mist holy. And it is true that our contemplation of the natural world does not give us a feeling of detached admiration, it makes us long obscurely for recognition: we wish to be contemplated in our turn.

I feel this too. In past ages, unshielded by urban noise and the iron grip of the scientific mindset, I think humans did feel the conscious embrace of the rest of the world and when they conceived the idea of a divinity " behind" nature, they sometimes heard the answering voice.

We are so unused to listening to the natural world, so full of the imagined solitude of our human consciousness, that we cannot believe that there may be an answering voice, here and now, from a plant or an animal who shares the living moment with us. And if an answer comes, we do not recognise it - it seems an absurdity, a fantasy, a New Age aberration to give such a voice any credence.

But my experience tells me that the eye of wild nature is watching us, there is an absolutely personal relationship, nature indeed possesses intentionality and is contemplating us while we contemplate her. But to discover this, we have to suspend our cynical disbelief and have the patience to observe the flow of things and then reflect on them at length, so as to see how things hang together. This requires the sort of life in which there is time and silence; country work, repetitive and rhythmic, is helpful for this process. It allows that receptive state of mind which is not sleep or dreaming, but which is an empty attentiveness.

In order to remove the veil, the first thing to do is to try and describe our own experiences, at least to ourselves, without censorship. Marco Chiletti, who farms in Tuscany, told me how he was once teaching a friend to prune olive-trees. He had given his friend a series of technical explanations for the cuts he had made, but his friend, after watching for a while, pointed out that some of the cuts Marco was doing contradicted the instructions he had given. Marco wondered why he had been talking in that impersonal way, like a manual, instead of giving explanations that were closer to his real experience. It is a habit, a way to avoid sounding ridiculous, even with someone we know well. Finally he managed to say "I have lived alongside this tree for many years, I cut in the places that the tree itself suggests to me: it is a conversation between us". If we could talk in these terms, it would be easier for us to recognise the reality we live in and easier for us to trust our intuition.

Last spring, I was down in a distant field where there were piles of juniper trees cut during the previous winter to keep the pastures clear. Martin had stacked the big logs for carrying up to the woodpile and he had discarded the smaller branches. It seemed to me a shame not to use that perfumed, resinous wood which burns so well when it's dry, so I decided to make the bundles of cooking wood down there. It was a beautiful, quiet place. The only sounds were from the waterfall cascading down the opposite hillside into the stream Cerquetelle and the mules belonging to a neighbour called Renato, who were carrying logs up from the steep wood across the valley. Every so often the sheep would come by, or the donkeys, and the dogs kept me company. I spent all the morning hours down there for several weeks.

Martin was furious. We had at least seven separate arguments over that job. In his opinion, the work down there was finished, I was interfering, this was a spot too far from home for carting all the bundles, there was plenty of small wood much nearer home and there were more urgent jobs to be done. In reply to all these good reasons I could only say that I really wanted to make these juniper bundles, it was special wood, it felt like a question of respect to use it, that I would make sure I carried the bundles up myself with the donkey and that the place was beautiful. Even as I argued, I was surprised at my obstinate stand against Martin's perfectly valid objections: I just had a strong feeling that I should do this job and I hadn't much in the way of logic to offer. In the end, I said "Look, could you just pretend I'm ill or gone off somewhere? I really want to do this job and I'm going to do it."

So for six weeks I immersed myself in the shapes of the trunks and branches of the juniper trees. The needles of the cut branches were still green...Sardinian mythology says that the spirit of the Juniper is so strong that it stays in the tree for a long time after it is cut. Every branch I added to the bundles felt like something valuable saved. There was a strong smell of resin in the air, and when the spring sun became hot, I was glad of the shade of the tall junipers that were still standing. As I cut the branches to the right length, I became intimately familiar with their forms, the way the blue berries cluster at certain points on the branch, the curves, the weak points and the tough places of the plant. By the end of the job, I felt I knew the juniper in that close, wordless way in which children know the flowers they have played with, a knowledge of hand and eye which has nothing to do with Botany.

Martin used to make lunch and when he called, I had to hurry up the steep slope to get home and I noticed that, as the days went by, the asthma that has always been a companion of mine over the last forty years was much better: I could breathe freely. The hillside seemed less and less steep. I thought it must have something to do with the many hours of quiet solitude and it was only months later that, leafing through a book of herbal remedies, I read that Juniper is a cure for asthma. But I hadn't eaten the berries...could my presence near the juniper trees have been enough? Then I remembered my stubborn, irrational insistence on doing that piece of work and in the end, I came to the conclusion that, even if it seemed absurd, it had been the junipers themselves "calling out" to me to come and do that piece of reciprocal care and attention. I felt I had been seen and embraced in precisely the same way that I had gathered up the juniper wood and treated it as something valuable. Of course, it is easy to laugh at this idea ("It's only this, it's only that"), but I made a conscious decision to trust the feelings that I had because I see that the desacralization of nature has not brought us good fruits. Perhaps we should try to trust our intimate experiences and see whether, if we give them the benefit of the doubt, they may lead to other, unexpected conversations of this sort.

In response to our question *for whose sake* the natural world operates, Plotinus attributed this reply to Nature: "It is not fit you should interrogate me, but it becomes you to understand in silence, even as I am silent, and not accustomed to speak: but what is it you should understand? this, in the first place, that whatever is produced is my spectacle, produced while I am silent, a spectacle naturally produced, and that I, who spring from a certain contemplation of this kind, possess a nature desirous of beholding...the lines of bodies emanate from hence, not by engraving them in matter, but drop, as it were, from the energy of my contemplation." Although Plotinus conceived nature as a "lower" manifestation in the scheme of things and matter as "non-being", this description of the arising of forms reminds us of Giordano Bruno's beautiful image of matter producing forms from out of herself just like the head and horns of a little snail appear slowly, slowly before our eyes. Bruno said of the Egyptians "those wise men knew that God is in all things, that the divinity is hidden in nature, shining and working through different agents, in all the various physical forms." (It is interesting to remember that Plotinus was Egyptian by birth).

What happens when we, who are one of the forms of nature, turn back and contemplate her? It is the beginning of the journey home and this is the moment which is symbolized by the winter solstice.

In the churches and cathedrals of Spain, France, England and Germany between the years 1000 and 1500, a figure, which English folklore calls the Green Man, began to appear in the decorative sculptures. It was actually a reappearance, because the Green Man has his origins in the Son-Lover of the Great Goddess of ancient Europe, the "Masked Ittiphallic God" as the archaeologist Marija Gimbutas calls him, the God who is sacrificed, buried and resurrected, the

God of vegetative life, the angel of the holy tree which appears in many sacred traditions, from the Vedas to the Nordic myths, and in general, all over pre-Christian Europe, where the religion was a religion of place and the place was the forest. He appeared in the Domus Aurea in Rome, Nero's "Golden House", in the wall-paintings described by Vitruvius that showed "plants with small human figures sitting amongst the fronds". It is very curious to imagine how the Green Man managed to survive at the margins of Christian culture, and appear in Romanic, Gothic and Renaissance art despite the Christian repression of all cults centred on holy trees, a process which continued until well into the 13th century. If you look carefully at the stone sculptures on the cathedrals of Chartres, Le Mans, Auxerre, Freiburg im Breisgau, Ulm, Bamberg, Ely, Exeter, Norwich, San Pietro in Toscania in Viterbo and S. Maria del Mar in Barcellona, to mention only some, you can see hundreds of Green Man faces, made of leaves, with leaves emerging from his eyes or mouth: an ancient imagine of the union of humanity with the world of vegetation, an image of the exuberance of life, renewal and rebirth. The Green Man knows the secrets of nature, the leaves which sprout from his mouth are the song with which the tree spirits speak to human-kind: they speak of the intelligence and intentionality of nature. The Green Man is the guardian and revealer of the silent mysteries. He is the image with which human beings have expressed their intimate experience of the green world. His is the face of the world soul.

Two studies of the Green Man have been made in the last few years: one by Kathleen Basford in 1978 (*The Green Man*, Ipswich) and another (*Green Man*, Harper Collins) by William Anderson and Clive Hicks in 1990. It is interesting how this figure is re-emerging from the shadows now, at a time when the Goddess has come to the fore of human consciousness in the West. This man of the Forests is a strong image for those men who are trying to forge their relationship with nature in the old, close way, letting go of images of conquest. Surely it can't be that the Goddess is the only positive face for the new embrace between humans and the natural world when that embrace needs a fertile, joyful union between men and women. In the church of St. Bertrand de Comminges there is a wooden carving of the Great Mother giving birth to the Green Man, a very rare image that shows us their deep connection. These images can only have survived the purging process as decorative framing for the central Christian images, but I wonder what meaning they had for the artisans who actually carved the figures. It is possible that pagan currents still flowed close below the surface of their lives, despite their outward allegiance to Christianity.

John Scotus Erigena, a medieval theologist, talked of the Garden of Eden as a symbol for human nature and the Tree of Life as the word of God, recapturing the numinous light around the sacred tree of tradition. Terence McKenna changed the emphasis in this triad of God-Man-Plant: he suggested that it may have been the meeting of early primate-humans with hallucinogenic plants which offered our ancestors the chance to communicate with the "spirit" behind the plant world, with the anima mundi or inner dimension of the world, and that this dramatic and first-hand experience of God "behind" the physical world was the trigger which catapulted humanity out of a purely animal life and into self-consciousness. If we conceive of God as entirely separate from creation, this may seem blasphemous to some, but if, like Giordano Bruno and many native peoples all over the world, we see creation as a manifestation, an outpouring of the deity, it is an entirely understandable idea.

William Anderson in his book asks himself "How do we speak to an archetype? Or rather, what do we have to do to get the archetype to speak to us?" and his reply is "Let the image grow inside you, return to it many times with the humble gifts of attention and silence."

I think the great secret is *attention.* A few months ago, I suddenly realised that the occasional episodes of horrible sickness and migraine I was suffering actually came at regular intervals and lasted a couple of days just before a menstrual period. "Oh, well, this will be the menopause" I thought, but I was alarmed: I had never suffered over periods or pregnancies or childbirth and I was upset at the idea of not riding smoothly through the next stage. Nowadays we are lucky enough to have a real doctor here and he gave me tincture of sage. The first month, I took the sage drops but I also went every day to the sage plant outside the courtyard gate and bowed to it and ate a leaf. The critical day came and I was a lot less ill. The next month, I tried not taking the drops, but eating a leaf and bowing to the plant every day and I was even better. The third month, in trepidation, I didn't touch the sage, but only bowed to the plant every morning and that month I was fine. Ask for help and it will come: the footprints of the Green Man are enough to show his presence.

The West is still suspicious of homeopathy and its infinitesimal quantities of natural substances, but the idea of "pure information" passing from a plant to a human being is part of the shamanic tradition. An Amazonian healer explained, "When I need to make medicine, I go and look for the plant that I know can help and I sit by the plant for a long time, in silence. I listen to the *plant's song* and when I have learnt it, I go home. I take a bowl of river water and then, holding the plant in my heart, I sing its song over the surface of the water. The water is the medicine. I do this because not all people are capable of listening for such a long time."

Civilizations who have lost such close contact with the plant world have given the plant-man connection a human face made of leaves because it is easier for us to perceive intelligence in a human face than in a green plant.

The Green Man does not loom only in the European mind: in the Islamic world, there is a figure known as Kidhr. The name means "The Green One" and Kidhr stands mysteriously at the gateway between the visible and the invisible worlds. His foot leaves a green footprint. He shines like light on the green grass, he appears as an inner experience, like the passionate capacity to work beyond human limits when inspired. Kidhr is an inner dimension that transcends form. He appears to the solitary man, the man who is cut off from the conventional religious traditions of mankind. Ibrahim, the king who gave up his kingdom to follow the life of a Sufi mystic, said of Kidhr, "I lived for four years in the desert. Kidhr, the Ancient Green One, was my companion. He taught me the Great Name of God."

Rumi, the 13th century Sufi poet and contemporary of Saint Francis of Assisi, speaks of Kidhr in his poem *Love Dogs.* Here is an insight into Nature's silent language:

One night a man was crying,
Allah! Allah!
His lips grew sweet with the praising
until a cynic said,
"So! I have heard you
calling out but have you ever
gotten any response?"

The man had no answer to that.
He quit praying and fell into a confused sleep.

He dreamed he saw Kidhr, the guide of souls,
in a thick, green foliage.

"Why did you stop praising?"
"Because I never heard anything back."

"This longing
you express *is* the return message.

The grief you cry out from
draws you toward union.

Your pure sadness
that wants help
is the secret cup.

Listen to the moan of a dog for its master.
That whining is the connection.

There are love dogs
no one knows the names of.

Give your life
to be one of them.

There are deep human needs, with and without names, and perhaps there is also a desire on the part of the rest of creation to be seen and understood by a contemplative human eye that perceives the intelligence (what used to be called wisdom!) in the natural world. We can't easily bring ourselves to believe that the Divinities may be so close, so everyday, so down to earth: we are used to sky Gods, distant, unreachable, unthinkable.

I spent some weeks last spring digging the roots of the yellow thistles out of the top pasture: Carlina vulgaris, "the most beautiful flower of summer" as Hesiod called it (this tells us Hesiod was not a peasant!) There were so many of them and it's no use in the long run cutting them back, you have to get the roots out but their roots are deep, often fifty centimetres long. After a few days, I had to stop because I had a thumb badly swollen up from a thistle spike. During the pause, as I mused on the huge area of thistles still to be dug up, I suddenly realised what an elegant system for the return of the forest cover these thistles are part of. They spring up quickly and, unchecked, in a year or two they can cover a field in a thin layer of prickles so that, although the pasture still looks viable to the casual glance, in fact the grazing animals won't eat there any more. In this way, the thistles have made the field safe for sloe-berry, blackberry, dog-rose and juniper, all thornily self-protective, and if these plants survive for a few years, tough-rooted broom appears and soon the field is full of maple saplings, oak, ash and hornbeam, and a small wood is growing. From the point of view of Nature, with her desire for climax, the thistles are a sophisticated initial response to a wound. And if one looks at them from a non-human point of view, not as a weed, they are indeed very beautiful. I could have come to this conclusion without plunging a prickle into my thumb, but the fact of the matter is that the sudden revelation of the perfection of those thistles came to me during the few days when I stopped attacking them, as if the Divinity of the plant had intimated "Cut them, but open your eyes to their beauty and perfection, look at them with understanding!"

It is strange how the Green Man seems to request silence before the workings of nature, and yet the green foliage which pours from his mouth promises words, song, inspiration. Perhaps he is asking for respect and attention for the green world, and if the human response is the silence of listening, miracles can happen, even the miracle of human truth-speaking. If we listen, we are listened to. If we listen, we can prophesy. There is an old pact between the human world and the vegetable world which we need to keep and which, if it is intact, opens our eyes to the inner life of the planet and our ears to its songs.

The poet Henry Vaughan (1622-1695) wrote a letter, replying to his cousin John Aubrey, who had asked him questions about the Welsh bardic tradition. " I was told by a very sober and knowing person (now dead) that in his times there was a young lad fatherless and motherless, and so very poor that he was forced to beg; but at last he was taken up by a rich man that kept a great stock of sheep upon the mountains not far from the place where I now dwell, who clothed him and sent him into the mountains to keep his sheep. There, in summertime, following the sheep and looking to their lambs, he fell into a deep sleep, in which he dreamed that he saw a beautiful young man with a garland of green leaves upon his head and a hawk upon his fist, with a quiver full of arrows at his back, coming towards him, whistling several measures or tunes all the way, and at last let the hawk fly at him which he dreamed got into his mouth and inward parts, and suddenly awaked in a great fear and consternation, but possessed with such a vein, or gift of poetry, that he left the sheep and went about the Country, making songs on all occasions, and came to be the most famous Bard in all the Country in his time".

Watch for the green footprint, hold close your inner experience, understand in silence.

BIOREGIONAL THINKING sees human beings as part of the wild community of all beings, woven into a web of relationships with the plant and animal life of a particular place. Each in our own place, we can share the work of living together and find delight in this ancient interdependence. Awareness of a shared place demands that we leave enough space, food and water for all our relations, human and non-human. If we can contain ourselves within these limits, our lives are strengthened and we can live the luminous dream of the earth. "Nature is not a place to visit, it's home."

Bibliography

Anderson, William and Hicks, Clive, *Green Man*, Harper Collins, London and San Francisco, 1990.

Baltvilks, Janis, *Conversation beneath the Trees*, Nordik, Riga, 1999.

Basford, Kathleen, *The Green Man*, Ipswich 1978.

Berry, Wendell, *The Selected Poems of Wendell Berry*, Counterpoint, Washington D.C., 1998.

Berry, Wendell, *Clearing*, Harcourt Brace Jovanovich, 1974.

Cardano, Mario, *Lo specchio, la rosa e il loto. Uno studio sulla sacralizzazione della natura.* Ed. Seam, Roma, 1997.

Falassi, A., *Folklore by the Fireside,* University of Texas Press, Austin, Texas, 1980

Fideler, David R., ed. *Alexandria, The Journal of the Western Cosmological Traditions,* Vol.I, Phanes Press, Grand Rapids, 1991.

Gifford, Terry, *Green Voices. Understanding contemporary nature poetry,* Manchester University Press, Manchester, 1995.
Hughes, Ted, *Moortown*, Faber and Faber, London, 1979.
Hughes, Ted, *Selected Poems 1957-81*, Faber and Faber, London, 1982.
Isaacs, Jennifer, *Australian Dreaming: 40.000 Years of Aboriginal History,* Ure Smith Press, Sydney, 1991.
Kavanagh, Patrick, *Collected Poems,* Martin Brian and O'Keefe, London, 1972.
LaChapelle, Dolores, *Deep Powder Snow,* Kivaki Press, 1993.
Lawless, Gary, *First Sight of Land,* Blackberry Books, Nobleboro, 1990.
Lawless, Gary, *Caribouddhism*, Blackberry Books, Nobleboro, 1998.
Maloberti, Francesco, *L'Abbadia e il Castello di Vallingegno*, Faglioni, Milano, *1967*
McKenna, Terence, *Food of the Gods*, Bantam Books, 1992.
Mumford, Lewis, *The City in History,* Penguin, London, 1961.
O'Brien, Flann, *At-Swim-Two-Birds,* Penguin Books, 1980.
Reynolds, Reginald, *The Wisdom of John Woolman,* Quaker Home Service, London, 1988.
Rughi, Luigi, *Memorie ss l'Abbadia di S. Verecondo in Vallingengo,* Oderisi, Gubbio,1935
Rumi Jalalad Din, *The Essential Rumi,* Transl. by Coleman Barks, Penguin Books, London, 1995.
Sakaki, Nanao, *Break the Mirror,* North Point Press, 1987/Blackberry Books, Nobleboro, 1996.
Sakaki, Nanao, *Let's Eat Stars,* Blackberry Books, Nobleboro, 1997.
Sheldrake, Rupert, *A New Science of Life,* Inner Traditions International, 1999.
Snyder, Gary, *Turtle Island,* New Directions, New York, 1974.

Snyder, Gary, *The Real Work: Interviews and Talks 1964-1970* , New Directions, New York, 1980.

Snyder, Gary *The Practice of the Wild,* North Point, San Francisco, 1990.

Strehlow, T.G.H. *Songs of Central Australia,* Angus and Robertson, Sydney, 1970.

Strehlow Research Centre, P.O.Box 831, Alice Springs, Northern Territory, 0871 Australia. Fax 089-518050. Study centre on the work of TGH Strehlow.

The Little Flowers of St. Francis, Edizioni Porziuncola, Assisi, 1982.

The Trumpeter, http://trumpeter.athabascau.ca, Athabasca University, Canada. Journal of Ecosophy (for article by Stan Rowe).

Thomas, D.S., *Later Poems,* Hart-Davis, London, 1966.

Thoreau, Henry D., *Walden or, Life in the Woods,* Penguin Books Inc, New York, 1942.

Tributsch, Helmut, *When the Snakes Awake: Animals and Earthquake Prediction,* MIT Press, Cambridge, Ma., 1984.

Waters, Frank and Oswald White Bear Fredericks, *The Book of the Hopi,* Penguin, Putnam, 1977.

Italian journals, publications and information sources in tune with the bioregional vision

Giuseppe Moretti, Reference person for the Italian Bioregional Network:
Rete Bioregionale Italiana,
Publishes the journal **Lato Selvatico.**
Address: via Digagnola 24, 46020 Portiolo, MN, Bioregione Bacino Fluviale del Po, Italy.
email:morettig@iol.it
Telephone: 0376-611265

Gaia Newsletter, Stefano Panzarasa, address: via dei Portici 39, 00010 Moricone (Roma), Italy.
Newsletter for the Tiber Valley Bioregion: Bioregione Bassa Valle del Tevere.
email: bassavalledeltevere@inwind.it
Telephone 0774.605084.

Bullettin, Paolo D'Arpini, address: Piazza Roma, Calcata, VT. Bioregione Tuscia, Italy.
Newsletter for the Tuscia Bioregion.
Telephone:0761/587200

Bioregione Valle del Savio
Wilderness experts: Maurizio Castellucci and Elena Avanzini.
Address: via Staffette Partigiane 220, 47020 Gattolino di Cesena, Fo, Bioregione Valle del Savio, Italy.
Telephone: 0547-324614

La Raccolta, Gino Sansone, vegetarian restaurant and cultural centre, address: Arcobaleno Fiammeggiante, c.p. 82, 80100 Napoli, Bioregione Partenopea, Italy.
Newsletter for the Neapolitan area.

AamTerra Nuova, Address: via Ponte di Mezza 1, 50127 Firenze, Italy.
email: info@aamterranuova.it
Monthly magazine on wholefoods, environmental issues and medicine.

CIR (Corrispondenze ed Informazioni Rurali)
Renato Pontiroli, address: Borgo Cerri, 17030 Erli, SV, Italy.
Telephone: 349-1578435
Newsletter put out by reinhabitants of the Italian bioregions

Seminasogni, Address: Loc. Palombara di Castel San Pietro 9, San Severino Marche, MC, Italy.
Newsletter published by the reinhabitants of the Marche region and Central Italian Appennines.

Native American Library of Perugia, address: via Fratti 18, Perugia, Italy.
Telephone:075/5720716
Hours of opening: Tuesdays and Fridays 101m - 3pm and 4-7pm.
Very well supplied with texts.

Fondazione Archivio Diaristico (Italian Archive of Personal Diaries)
Address: Piazza Plinio Pellegrini 1, 52036 Pieve Santo Stefano AR, Italy.
Telephone: 0575-797730 email: adn@archiviodiari

$14.00

Wild boar, oak, flowering ash, deer , dogwood, badger, dogrose, porcupine, juniper, cuckoo, milk thistle, buzzard, sloeberry, crow, and the donkey Otello, the olive trees, the horse Raja, the apple trees, the Sardinian sheep, the mulberry in the courtyard, the doves, the yellow plum trees, the hens and ducks…they are the non-human neighbours we live with on this hillside and they fill our days with delight and disaster. Who would want to live without them?

9

